Dayse Suellen Moraes
Manoel F. Gomes Filho

Contribution of Rainfall in the Afternoon in Belém, Pará

Dayse Suellen Moraes
Manoel F. Gomes Filho

Contribution of Rainfall in the Afternoon in Belém, Pará

Study on the Contribution of Rainfall to Monthly Totals and Possible Relations with the Climatological Average

ScienciaScripts

Imprint
Any brand names and product names mentioned in this book are subject to trademark, brand or patent protection and are trademarks or registered trademarks of their respective holders. The use of brand names, product names, common names, trade names, product descriptions etc. even without a particular marking in this work is in no way to be construed to mean that such names may be regarded as unrestricted in respect of trademark and brand protection legislation and could thus be used by anyone.

Cover image: www.ingimage.com

This book is a translation from the original published under ISBN 978-613-9-70021-9.

Publisher:
Sciencia Scripts
is a trademark of
Dodo Books Indian Ocean Ltd. and OmniScriptum S.R.L publishing group

120 High Road, East Finchley, London, N2 9ED, United Kingdom
Str. Armeneasca 28/1, office 1, Chisinau MD-2012, Republic of Moldova, Europe
Printed at: see last page
ISBN: 978-620-8-15023-5

I dedicate this work to my grandfather, Inocência Brito de Moraes (In Memoriam), who supported me unconditionally at all times, especially in times of uncertainty, which is very common for those trying to tread new paths.

ACKNOWLEDGEMENTS

Coming to Campina Grande to study for a master's degree was a decision marked by the desire to grow professionally and, at the same time, try to give my personal life a new lease of life. Being away from my family, my friends, in short, from where I live, was not an easy task, but it was also one of the best things life has taught me. Allied to this arduous side of my experience, there were great compensations, such as the arrival of new friends, people I could count on in various situations, even the most complicated ones, and who made me feel that, even away from my place, I wasn't alone.

That's why I'm grateful:

Firstly to God for all the blessings he has given me, for his constant presence in my life, lighting my way, giving me the strength to follow him and for the opportunity to achieve yet another goal;

To my family, the foundation of everything I am and for allowing me to make important decisions that led me to this achievement, for their personal and professional guidance, daily encouragement, support in difficult times, unconditional love and for believing in my ability;

The Federal University of Campina Grande (UFCG) and the Postgraduate Programme in Meteorology, for the opportunity to take this course.

To Dr Manoel F. Gomes Filho for his guidance, patience and support in carrying out this work and for his friendship over the years;

To the professors of the Postgraduate Programme in Meteorology - PPGMet/UACA/UFCG for their teachings during my master's degree, further increasing my appreciation for this institution;

To the 2° Meteorological District (2° DISME) for providing the information needed to carry out this research;

To Lady Layana Martins Custódio, a great friend and walking companion throughout these two great years. I can say that this journey, including my personal one, would not have been the same without you. I thank you for your encouragement, your help with the development of this work, your patience and your affection, respect and exchange of academic information;

To my friends, Waléria Figueira Souza, Pâmela Ávila and Leydson Galvíncio, for their friendship as soon as I arrived in the city of Campina Grande - PB, as well as their support throughout this period;

To my fellow master's students, whom I had the pleasure of getting to know, especially Danielle Cagliari de Lima, for the camaraderie and the exchange of experiences when necessary;

Finally, there were many people who helped me, directly and indirectly, in the realisation of this work, and it would not be possible to mention them all here, but I would like to express my deep gratitude to all of them;

And last but not least, to the Coordination for the Improvement of Higher Education Personnel (CAPES), for the financial support during the development of the research project.

I can do all things through him who strengthens me. "

(Philippians 4-13)

SUMMARY

The contributions of afternoon rainfall in Belém and the homogeneous rainfall regions were determined using statistical analyses and clustering. Hourly and monthly data from 25 rainfall stations distributed throughout the state of Pará were used. In general, the results showed that afternoon rainfall in Belém contributes an average of 17 per cent of the annual total. The hourly cycle of annual rainfall showed that the contributions are different for each month of the year. In the less rainy months, this percentage is around 30 per cent. Precipitation rates showed maximum peaks at 4pm local time, directly influenced by the action of the ITCZ, the main mechanism generating rainfall in Belém, and by LI and local systems. It is suggested that the inter-annual oscillations in seasonal rainfall are mainly due to climate variability associated with the Southern Oscillation and Atlantic Dipole index events. The cluster analysis technique applied divided Pará into three homogeneous rainfall regions: Group 1 - located in the north-west, with rainfall mainly influenced by the north-south displacement of the ITCZ; Group 2 - located in the north-eastern part of the state, where local convection, MLF, LI and the ITCZ are the main atmospheric systems producing rain; and Group 3 - located in the centre-south of the region, where rainfall is subject to the influence of frontal systems along with local convection and the ITCZ. Group 2 was analysed, where municipalities with rainfall similar to the city of Belém were identified. The hourly data from these stations was then analysed and the results obtained were that Belém has a rainfall regime that differs from the other municipalities, being the only municipality to have an increase in rainfall in the afternoon in all months.

Keywords: Belém, Rainfall, Contribution, Hourly cycle, Group, ENOS, ITCZ.

Summary

CHAPTER 1

INTRODUCTION

The city of Belém, capital of the state of Pará, is characterised by high rainfall rates due to its proximity to the equator. The region's climate is influenced by mesoscale atmospheric systems (such as sea breezes and lines of instability) (KOUSKY and MOLION, 1981; COHEN, 1989), synoptic scale systems (easterly wave disturbances and high-level cyclonic vortices) (MOTA, 1997; ASNANI, 1985; LIEBMANN, 1990; NITTA et al, 1985; YAMAZAKI, 1975) and large-scale (Intertropical Convergence Zone) (KOUSKY and KAGANO, 1981; NIMER, 1989; SANTOS, 2006; CAVALCANTE et al., 2009) that can cause large volumes of rain (FIGUEREDO and MOTA, 2010).

Belém has a rainy season from December to May, with the highest rainfall in March, influenced by the Intertropical Convergence Zone (ITCZ), and a less rainy period from June to November, as described by Nechet (1984a), Bastos et al. (2002) and Oliveira et al, (2003), with the month of November having the least rainfall, which is caused by Lines of Instability (LI) that form in the coastal part of the state, due to the circulation of the Sea Breeze and head inland towards the continent as lines of convective clouds, causing rainfall mainly during the less rainy season in the region (COHEN et al., 1995).

In Belém, the traditional question is: "Are you going out before or after the rain?". Those who don't know Belém may be startled by this question, but it's quite common in the region due to the hot and humid climate. However, it's worth pointing out that, on average, it doesn't rain in Belém on approximately 30 per cent of the days of the year, and 40 per cent of the days it rains between 0 and 0.9 mm/day. It rains practically every day in the city and almost on schedule (at 3pm). However, this rainfall has been jeopardised, especially during the Amazonian summer (September to November) when there is usually little rainfall, coupled with other climatic phenomena such as the EI Nino.

It is also important to mention the ocean-atmosphere climate mechanisms that indirectly influence precipitation in the region. The best known is the Nino-Southern Oscillation, which is the result of an interannual fluctuation in atmospheric pressure at mean sea level between the Western Equatorial Pacific and the Southeastern Subtropical Pacific that alters atmospheric circulation, which in turn modifies the Sea Surface Temperature field along the Equatorial Pacific, alternating between positive (warm) and negative (cold) anomalies.

The other is the Tropical Atlantic Meridional Sea Surface Temperature Gradient, which consists of a simultaneous manifestation of sea surface temperature anomalies with opposite signs in the northern and southern tropical Atlantic Ocean basin. This configuration generates a shift of the ITCZ towards the hemisphere with the highest ocean surface temperature values, thus influencing the rainfall regime in the Amazon (DE SOUZA et al., 2004).

Even though it's common to imagine that the rain stays at the same time, statistically such punctuality is not observed. The rain is no longer timed. There is variability in the times, seasonal variability. Nechet (1984b, 1997) observed that during the wettest period, from the second half of December to mid-May, rainfall normally occurs between 2 and 3 pm. After this peak, there is a shift to later in the day, i.e. at the end of the rainy season (June, July, August) between 5 and 9 pm, and between September and December, it gradually begins to return to the middle of the afternoon, between 3 and 4 pm.

Studying the spatio-temporal variability of rainfall in a region is fundamental for planning its activities. The identification of rainfall patterns and homogeneous rainfall regions has been a common practice and has been developed using different methodologies. The Cluster Analysis technique has shown promising results in terms of more consistent regionalisation of local climatic rainfall characteristics (GREEN et.al., 1993, GONG and RICHMAN, 1995, MUNOZ- DIAZ and RODRIGO, 2004, FRAGOSO and GOMES, 2008, STATHIS and MYRONIDIS, 2009, GRIMM, 2011), providing a better understanding of the patterns of spatio-temporal variability of the data, which are often not evident when working only with their average or seasonal values (AMANAJAS and BRAGA, 2012). Although rainfall analyses for the Belém region, which is characterised by high rainfall and no dry period, have been carried out in recent years, for example by Loureiro (2014), there is a lack of studies for this region.

Therefore, the focus of this book is to provide unprecedented information on the behaviour of rainfall on an hourly time scale and to quantify the importance of Belém do

Pará's "famous afternoon rain" in the average monthly rainfall total, since there are few studies that focus on the variation of rainfall on an hourly scale, and to date there have been no publications using automatic stations with long historical series for the region.

1.2. OBJECTIVES

1.2.1. General Objective

The general aim of this book is to qualitatively and quantitatively assess the contribution of afternoon rainfall in Belém in relation to the daily total from 2003 to 2015 and the possible relationship with the average for the entire state of Pará.

1.2.2. Specific objectives

a) Analyse the percentage contributions;

b) Using Ward's method to map homogeneous rainfall in the state of Pará;

c) To identify the physical factors within the state of Pará that influence the spatialisation of rainfall;

d) Identify regions in the state of Pará with rainfall similar to Belém in the afternoon.

CHAPTER 2

LITERATURE REVIEW

The Amazon region is characterised by a year-round hot and humid climate and rainfall is the parameter with the greatest heterogeneity and spatial-temporal variability. This region has two seasons which are divided into: the rainy season (from February to May) and the less rainy season (from September to November); the other months of the year are considered transitional (LOUREIRO et aL, 2014).

The main atmospheric systems that act on the weather and climate of the Amazon region are: the ITCZ, SACZ (South Atlantic Convergence Zone), LI (Lines of Instability), MCC (Mesoscale Convective Complexes), DOL (Eastern Wave Disturbances) and the BT (Land and Sea Breezes) and BF (River Breezes) (LOUREIRO et aL, 2014; FISCH et aL, 2010).

Below we will describe the main meteorological systems that act to directly or indirectly influence rainfall in the afternoon in the state of Pará and specifically in the city of Belém.

2.1. Main meteorological systems operating in the state of Pará.

The main meteorological systems operating in the state of Pará include the following.

2.1.1. Sea and Land Breezes

Identified by the ancient Greeks since around 350 BC and representing a local wind system (VIANELLO and ALVES, 1991), or a specific wind regime, on a diurnal scale, resulting from thermal differences between land and sea surfaces that lead to a difference in air pressure at low levels. Over these surfaces, breezes are of two types: sea breezes, called BM, and land breezes, called BT. They are induced by spatial heterogeneities of the surface heat flux in the planetary boundary layer (ROTUNNO et al., 1992). The penetration of the BM in North/Northeast Brazil (NNEB), studied by Kousky (1980), together with the valley-mountain circulations, determine the preferential times of occurrence of precipitation. The same author also showed that the seasonal variation in the direction of the trade winds (on a planetary scale), when coupled with the BT, produces a convergence zone (visible as an "arc" of clouds on a satellite image) over the ocean area close to the coast. In continental areas, the presence of a gradient wind can substantially affect the formation of the breeze cell or even hinder its penetration into the continent. If the wind is blowing against the coast, i.e. from the sea to the land, the temperature gradient decreases, which hinders the establishment of the cell, and likewise may not allow the formation of BT, as it is generally weak. In places with a weak offshore wind, the sea breeze forms earlier (08h00 HL) and moves 20-30 km inland by around 12h00 HL (ATKINSON, 1981).

Among the circulations, the strongest is usually the BM, which can move great distances, especially when it is formed with the help of topography. The BM near the coast starts to feel a wind coming from the sea in the late morning, which peaks in the early afternoon and disappears by nightfall. This wind is strongest on hot days, but can be weaker when the sky is cloudy. BT is still less understood than BM (ROTUNNO et al., 1992).

Machado et al. (2004) found that the BM dominates convective activity in Belém, contributing to the strong diurnal variation of convective clouds in the region. Although they are called secondary circulations in relation to the large scale, they are of great importance for the areas affected by it, as they contribute significantly to the total amount of precipitation and to the transport of pollutants (BAUMMBACH and VOGT, 1999). In addition, the breeze effect alters temperature, humidity and cloudiness (PLANCHON et al., 2006).

Studies on continental rainfall as a result of BM, lacustrine and mountain-valley influences are abundant in scientific circles, but studies showing BT influences on rainfall over continental areas are scarcer, not least because BT is generally expected to favour rainfall over the ocean and not over the continent (TEIXEIRA, 2008).

In the coastal region (from Pará to Amapá), precipitation is also high and there is no defined dry period, due to the influence of the LI that form along the coast in the afternoon and are forced by the BM (CLIMANÁLISE, 2010). Forming from 1800 UTC (1500 HL), surface heating and breeze circulation on the coast leaves the atmosphere favourable for the formation of clouds with high tops, increasing the likelihood of precipitation, especially in the form of showers.

According to Santos et al. (2012), when studying the effect of the BM in the city of

Belém, PA using multivariate analysis, they observed a daily shift in the local wind, both in the rainy season and in the less rainy season, varying from north-east to south-west, due to the effects of the BM circulations and realised that there is a dependence of the maximum rainfall on this horizontal wind shift, which is linked to the BM.

2.1.2. River Breezes

Oliveira and Fitzjarrald (1993) prove the existence of a fluvial circulation in the tropical region, at low levels (up to 1500-2000 m), in the forest/river direction during the night and early morning, reversing the direction (river/forest) during the afternoon and early evening.

Radiometric observations made by aeroplane during the ABLE experiment *(Amazon Boundary Layer Experiment)* observed a thermal gradient between river and forest of -3° C during the day and +6° C at night. These influences are certainly more intense in regions where the width of the river is considerable, such as near Manaus (confluence of the Negro and Solimões Rivers), Santarém (Tapajós and Amazonas Rivers) and Belém (Tocantins River and the southern part of the mouth of the Amazon River).

River breeze circulation is a characteristic noted in the large rivers of the Amazon, which occurs due to the different thermal capacity between the surface of the river and the surface of the ground being warmer than that of the river during the day, with upward air movement over the ground due to low atmospheric pressure and flow from the river to the land at low level (SILVA DIAS et al., 2004).

Amanajás and Braga (2012) identified, using Ward's agglomerative method, four homogeneous regions in the eastern Amazon and the region they identified as RH3 (which includes Belém), would have one of the main causes of rainfall, on a local scale, the BF.

In more recent studies, Matos and Cohen (2016) studied a convective system, observed for the first time in the Belém region during the CHUVA project campaign - Cloud processes of tHe main precipitation systems in brazil: A contribUtion to cloud resolVing modelling and to the gpm (globAl precipitation measurement) (MACHADO et al... 2014), 2014), which was called the Micro Fluvial Instability Line (MLF) due to its formation being associated with the BF circulation coming from Marajó Bay, which caused precipitation from its initial phase over the Belém region until it was over the western shore of Marajó Bay, increasing its intensity in this region.

2.1.3. Mesoscale Convective Systems

Among the main mechanisms generating precipitation over the Amazon basin are Mesoscale Convective Systems (MCS). These systems are formed mainly by Cbs clouds that agglomerate and organise themselves on the meteorological scale of time and space defined by Orlanski (1975) as mesoscale.

SCMs are often between 20 and 200 kilometres in length, but can reach dimensions of 200 to 2000 kilometres (MACHADO and ROSSOW, 1993). In addition to the large amounts of precipitation that these systems cause in the tropics and mid-latitudes, they represent the largest convective storms in existence (HOUZE, 2004) and are responsible for vertical energy transfers in the tropical troposphere (LAURENT et al., 2000).

Although they have similar general characteristics, SCMs are classified according to their shape (linear or circular), size, lifetime and cloud top temperature. LI, Mesoscale Convective Complexes, tropical convective clusters are some of these classifications (KOUKY, 1980; MADDOX, 1980; COHEN et al., 1995; VITORINO et al., 1997; SILVA DIAS et al., 2004; LU et al., 2005; FITZJARRALD et al., 2008; RAMOS DA SILVA et al., 2011; COHEN et al., 2014; MATOS and COHEN, 2015).

Maddox (1980) and Maddox (1983) found that the mesoscale organisation of Cbs occurs when convection is embedded in regions with a supply of warm, humid air. In the case of the Amazon basin, the supply of humid air comes mainly from the Tropical Atlantic Ocean transported by the trade winds (SATYAMURTY et al., 2012; DRUMOND et al., 2014).

Filho et al. (2015) studied convective precipitation based on integrated satellite data over South America with a spatial resolution of 8 km and a temporal resolution of one hour. The period used comprises 6.5 years and covers the equatorial region, between 0° and 15° s, which includes much of the Amazon, NEB and the adjacent Atlantic Ocean. As the study covered a non-homogeneous area (Amazon, NEB and ocean), the characteristics of the systems found could not be taken as pure characteristics of Amazonian SCMs. Furthermore, Amazonian systems with durations of more than 10 hours are not so common (VELASCO and FRITSCH, 1987; SALIO, et al., 2007).

When analysing the main characteristics of the behaviour of continental and oceanic

SCMs over the Amazon basin during the Green Ocean Amazon (GOAmazon) project, Rehbein (2016) found that the occurrence of below-average SCMs occurred distributively throughout the year and over the entire Amazon basin, and the possible causes were the strengthening of JBNs (COELHO et aL, 2015; SETH, et al...), 2015), the transport of moisture from the Amazon region to South America, which would imply a lower occurrence of SCM over the Tropical region (SALIO et al., 2007); ATSM above +0.5°C in the Eastern Equatorial Pacific Ocean in most months in 2014, accompanied by weakened trade winds in most months over the Atlantic and Pacific Oceans (GREC/IAG-USP, 2014) and the configuration of the EI Nino (EN) in 2015, which may cause suppression of Amazonian convection (ROPELEWSKI and HALPERT, 1987; MARENGO and HASTENRATH, 1993; GRIMM, 2003).

Also according to Rehbein (2016), SCMs show great spatial variability similar to the annual convection cycle, since during the summer the genesis of SCMs occurs spread over the entire Amazon territory and during the winter it moves to the north-east. Regions with a higher density of genesis stand out from the others and are located near higher topographics or at the confluences of large rivers or bodies of water.

In the warmer months, from October to April, they occur in the following favoured regions of genesis: on the slopes of the Andes, between northern Mato Grosso and southern Pará (in February this peak practically disappears and in April it is displaced to the east); on the Guiana Plateau (in October it doesn't appear); on Pico da Neblina and at the confluence of the Tapajós and Amazonas rivers.

Between May and September, the greatest activity in the genesis and development of SCMs shifts northwards and the favoured genesis regions become more restricted in this part of the Amazon. Between July and December there is another region of greater genesis, located on the coast of Pará, near the mouth of the Amazon River.

2.1.4. Mesoscale Convective Complex

The SCMs were called Mesoscale Convective Complexes (MCCs) by Maddox (1980), based on the analysis of dozens of cases using infrared satellite images over the mid-latitudes of the United States, who defined MCCs as a set of cold, thick Cbs with a circular shape and explosive vertical growth over a time interval of 6 to 12 hours.

The vast majority of studies dealing with the formation and development of MCCs, particularly in South America, have focused on those commonly observed in the mid-latitudes of northern Argentina, Paraguay and southern Brazil (GUEDES, 1985; VELASCO and FRITSCH, 1987; SCOLAR and FIGUEIREDO, 1990; ROCHA, 1992; DUQUIA and SILVA DIAS, 1994; MACHADO et al, 1994; GUEDES et al., 1994; FIGUEIREDO and SCOLAR, 1996; among others). The results of these studies indicate that subtropical MCCs occur preferentially during the spring and summer months of HS, forming at night, with a life cycle of between 10 and 20 hours. The meteorological systems that develop there have their dynamic and thermodynamic processes strongly influenced by the Andes mountain range.

The tropical region is not modulated by synoptic systems as in the subtropics, but it is associated with radiative heating of the surface which, in turn, promotes convergence and convection. This is why they develop most during the day. The Circular SCMs studied around the world (MADDOX, 1980; VELASCO and FRITSCH, 1987; VITORINO et al., 1997; SALIO et al., 2007) have their own organisation, intense convective activity, preferences for certain regions and times of day, and appear as a single isolated cloud cell with a rounded shape, which allows it to be easily identified from satellite images, and when it complies with the space-time and intensity classification made by Maddox (1980), it is called a CCM. These tropical systems have a lifespan of less than 9 hours, a size of less than 150,000 km2 during the period of maximum extension and are most frequent in summer followed by autumn (SALIO et al., 2007).

The Belém region also experiences the occurrence of SCMC-type mesoscale atmospheric systems (SODRÉ, 2013). These, like other mesoscale meteorological systems, depend on the characteristics of the local surface, such as topography, hydrography and vegetation, which directly influence convective processes, favouring or inhibiting their formation (ROCHA, 1992). However, the presence of these three factors in the same environment is not necessarily fundamental or decisive, since SCMCs can form in the presence of only one or even in the absence of all these characteristics (BARBOZA and FEDOROVA, 2000).

Sodré and Vitorino (2013) sought to identify SCMC, using remote sensing data, during ENOS climate extremes over the eastern Amazon, showing the interference of EN and/or La

Nina (LN) phenomena in the formation of circular mesoscale in the state of Pará. The authors observed that during the analysis the frequency of oscillation of the systems depends on the ENOS values, but the intensity of the volume of precipitation is controlled by the dynamic behaviour generated by these anomalous atmospheric mechanisms that modulate the climate. They concluded that in LN years the systems are more intense and in EN years they are less significant in terms of rainfall. And their frequency, which was basically similar in both EN and LN years, may not only be related to the large scale, but also to local factors such as topography and smaller scale mechanisms.

Sodré et al. (2015) carried out an observational analysis of SCMC to identify how different surface characteristics affect the genesis of these systems and found that the preferred time of formation of the systems depends directly on the characteristics of the surface, where, in flatter areas, SCMC tend to appear in the middle of the afternoon with a maximum peak at the beginning of the night, and in regions of steep relief the systems tend to form in the early hours of the night with a maximum peak at the beginning of dawn. In general, Amazonian SCMCs have an average lifespan of between 5 and 7 hours and the action of BM, lake breezes and valley-mountain breezes favours the transport of humidity, making it the main factor in the maintenance of SCMCs, especially in the northeast and northwest regions of Pará, where the humidity provided by the rivers and forest that predominate in these regions generates dynamic conditions favourable to the formation of SCMCs.

2.1.5. Lines of Instability

ILs in the tropics have been studied since the 1940s (HAMILTON and ARCHBOLD, 1945). With the execution of some field experiments, tropical ILs could be widely studied, as in Zipser (1969), Betts et al. (1976), Houze (1977), Cohen et al., (1989), Cohen et al., (1995), Silva Dias and Ferreira (1992), Alcântara et al. (2011) and Alcântara et al. (2014).

The first indication of LI in the Amazon region was given by Kousky (1980) in an observational study. This author suggested that the convection associated with the BM on the Atlantic coast enters the continent as a LI when the mean flow at 850 hPa is perpendicular to the coast and when the same flow is parallel to the coast the propagation of this LI does not occur.

Gamache and Houze (1982) define a tropical IL as the front region of a disturbance that propagates on the mesoscale. It is basically made up of Cumulonimbus clouds (Cbs) and in its rear by an anvil-shaped cloud which, in general, has a lot of stratiform precipitation. The convective elements that make up an IL are of the "hot tower" type (the same type as those discussed by RIEHL and MALKUS, 1958; and RIEHL and SIMPSON, 1979) containing upward flows that, by thrust, remove moist air from the boundary layer into the upper troposphere.

On the other hand, according to Zipser and Houze (1977), there are two types of downward flow: one on a convective scale, which occurs in a small band of 10 to 20 km in the region of intense precipitation on the LI itself, causing the gust front; while the other is gentler, but on a mesoscale, occurring in a large band of 100 to 500 km in the region below the anvil.

Cavalcanti (1982) carried out a climatological study and observed that the formation of these lines is positioned to the south of the ITCZ, with the period of greatest frequency being when the ITCZ is most organised. The climatological aspects of these lines in the Amazon were studied observationally by Cohen et al. (1989), who found that these LI can reach the far west of the Amazon, with a speed of movement between 12 and 15 m.s -1 (approximately 13° of longitude per day). The average length and width of these LI is approximately 1500 km and 170 km, respectively, and the most frequent months of occurrence are between April and August.

The ILs that occur in the Amazon are responsible for the formation of rainfall near the coast of the states of Pará and Amapá. Studies such as those by Cohen et al. (1989) have shown that these ILs are one of the atmospheric systems active in the eastern area of Pará and that they contribute 45% of the rain that falls during the rainy season. These lines are characterised by having large conglomerates of Cbs clouds and are formed due to the BM circulation. They can extend into the interior of the continent (called LIP) or remain confined to the coast (LIC). Due to their size, these ILs are easily observed by satellite images. According to the authors, LIC make up 62 per cent of the cases observed, with the remainder (38 per cent) comprising LIP.

Silva Dias and Ferreira (1992) classify the LI originating off the northern coast of Brazil, according to their inland propagation, into: (1) Coastal Convection Line (CLC), no further than 170 km from the coast; (2) Type 1 Line of Instability (SL1), moving between 170 and 400 km from the coast; (3) Type 2 Line of Instability (SL2), moving more than 400 km from the coast.

Cohen et al. (1995) studied LI penetration in the Central Amazon region during the ABLE-2B experiment (GARSTANG et al., 1990) in April-May 1987. During this period, LI comprised 23 per cent of the cases, the rest being IPL. In a case study of an intense LIP at the beginning of May, they observed that two mechanisms probably originated and propagated this LIP: intense easterly winds at low levels (between 900 and 650 hPa) and the presence of a heat source to the west of the Amazon. The authors conclude that the ILLs that propagate in the Amazon are complex mechanisms in which there is interaction between scales: large, meso and micro, the latter being caused by circulations between clouds and the surrounding environment.

Cohen (1996), using numerical simulations, found that the propagation of ILs formed along the coast is due to internal gravity waves propagating in a trough formed by the easterly jets near 700 hPa and the westerly jets at 200 hPa.

According to Gandú and Silva Dias (1998), humidity at low (700 hPa) and medium levels (500 hPa) and the flow at high levels (350 hPa) over the northern coast of Brazil are factors associated with the formation of LI. Moisture at low levels may be the main factor in the formation of IL because, in addition to supplying moisture to the system, it may be acting as a heat source that induces flow at high levels.

Barros (2008) showed that coastal convection from MCCs (isolated thunderstorms) can produce monthly precipitation accumulations of approximately 150 mm (40 mm) in the month of February (June). These systems that form in the coastal region but cannot be classified as LI are grouped by Oliveira and Oyama (2011) into a category called coastal convection.

Alcântara et al. (2011) assessed the distribution of the number of IL cases and verified the role of the Low Level Jets (LNJ) in the development of IL in the Amazon using GOES-10 satellite images from the Infrared and Water Vapour channel to identify IL that occurred between January 2000 and December 2008, following the methodology proposed by Cohen (1989). The LI can be identified in the images as a band of convective clouds, organised and aligned with the coast, between the hours of lOhOO UTC and 18h00 UTC. The JBNs were characterised by the identification of a relative speed maximum between the 900 and 600 hPa levels.

Alcântara et al. (2014), using the BRAMS model (FREITAS et al., 2007) version 5.02, studied the characteristics of the dynamics associated with ILs. Eight simulations were carried out, the main one called the control simulation which used sounding data taken from Wyoming University for Belém - PA at 12:00 UTC on 30 May 2008, which corresponds to moments before the formation of a type 1 propagation line, and the other simulations were carried out by changing the values of some components in order to understand the impact of wind shear at medium and low levels on the formation and development of the system, especially the contribution of the JBN.

Oliveira et al. (2016) analysed the wind profiles of the zonal (u) and meridional (v) components at a grid point closer to the city of Belém - PA at 12:00 UTC for days with and without the formation of LI, and found that the highest number of LI occurrences is seen between the months of April and August, which are also the months with the highest occurrence of BNW. This higher frequency corresponds to the period when the ITCZ is more organised near the coast.

The lowest occurrence would be in the months of October, November and December, confirming a possible relationship between the presence of JBN and the formation and development of LI.

2.1.6. Eastern Wave Disturbances

The discovery of Eastern Wave Disturbances (ELD) occurred in the Northern Hemisphere, probably with Dunn (1940), who noticed a shift in the 24-hour isalobars of a disturbance propagating from east to west in the Caribbean region, becoming a pioneer in studies on ELD along with Riehl (1945), who noted that changes in wind direction accompanied the formation and displacement of isalobaric centres and that these changes suggested that these centres were manifestations of atmospheric waves propagating from east to west. This type of atmospheric system was then defined as an easterly wave. This name was later changed to DOL.

Also known as African waves, easterly disturbances, easterly wind waves and wave disturbances in the trade winds (INNOCENTINI et al., 2005; RODRIGUES and CAVALCANTE, 2006; COUTINHO and FISCH, 2007 and TORRES, 2008), ELDs are systems that have been known for more than half a century and are considered to be one of the most important in terms of the precipitation regime in many tropical regions of the globe. They can be defined as synoptic disturbances associated with troughs and high sea surface temperatures (CHOU, 1990). Normally, LDOs are best configured at low levels (~850-700 hPa), where they have troughs, and their cloudiness is characterised by generally convective clouds, which move from east to west over the tropical oceans, reaching continental areas, such as the East of Northeast Brazil (LNEB) (ESPINOZA, 1996; VAREJAO-SILVA, 2005).

From the point of view of cloud cover, Kayano (1979) and Machado et al. (1997) demonstrated periodicities of wind and convective cloud cover with temporal frequencies related to DOL. It is hypothesised that DOL has a significant influence on the formation of LI formed off the coast of north-eastern Brazil (NNEB), which then spread inland to the Amazon. Kayano (1979), studying climate and circulation using radiosondes from Belém and Manaus, suggests that oscillations in the wind field with a periodicity of 3 to 5 days at low levels are related to the LODs affecting this region. In addition, LODs can intensify other convective systems such as ILs, for example (BARBOSA, 2005).

Barbosa et al. (2006) propose the following conceptual model for the formation of convective disturbances on the NNEB coast: given the formation of convective clouds, initiated by local factors (heating and breezes), if the convection is intensified by some convective synoptic-scale forcing, it will gain great horizontal extension and spread inland towards the Amazon. In this context, the authors suggest that the long-term convective disturbances correspond to the LI that penetrate the interior of the Amazon region and that some are forced by the DOL.

Torres (2008) studied the DOLs that affected the LNEB during the rainy periods of 2005 and 2006. During this period, the disturbances occurred in 26 events, with an average propagation speed of 10 m. s -1, a periodicity of five days and a wavelength of 4000 km. The author used satellite data and Hovmõller-type diagrams and found that the amplitude of the oscillations gradually increased from April to July, indicating the possible existence of a well-defined seasonal cycle for DOLs, i.e. a progressive increase in the number of DOLs over the

LNEB rainy season.

Oliveira et al. (2009) and Pontes da Silva (2011) observed in their analyses that the entry of cold fronts into the South Tropical Atlantic Ocean, which sometimes reach the NEB, and the presence of the ITCZ further south, affect the south-easterly trade winds in such a way that they end up converging with the southerly winds of the post-frontal high that accompanies the system, when both are intense in the hours preceding the intensification of the DOL, strengthening the transport of humidity from the ocean to the LNEB, contributing to the formation and intensification of the DOL that advances across the continent, causing rainfall in the NNEB and northern region (in this case, with its cloudiness already coupled to the ITCZ).

Caetano (2011) revealed that LODs occur mainly in winter and autumn and that the penetration of frontal systems from the HS to lower latitudes, associated with an increase in the gradient of ocean flow towards the continent, the presence of a cavado at low or medium levels of the atmosphere and SSTs above 27°C, represented a synoptic pattern very conducive to triggering or intensifying these disturbances.

2.1.7. Intertropical Convergence Zone

Among the large-scale meteorological phenomena, the ITCZ stands out, formed mainly by the confluence of the northeasterly trade winds from the Northern Hemisphere (NH) with the southeasterly trade winds from the Southern Hemisphere (SH). This phenomenon is characterised by a band of discontinuous cloudiness, as it is made up of distinct clusters of clouds, which are separated by regions of clear sky (MENDONÇA and DANNI-OLIVEIRA, 2007; UVO, 1989).

The intensity of the ITCZ is variable in both time and space, and precipitation is greatly exceeded due to the moisture provided by evaporation from the ocean surfaces below it and by evapotranspiration in some continental regions. Studies by Uvo (1989) and Waliser and Gautier (1993) show that the position and intensity of the ITCZ are closely related to the surface conditions of the ocean and atmosphere.

According to Marengo and Hastenrath (1993), rainfall in northern Brazil is associated with the seasonal fluctuation of the ITCZ. This system can act in two indirect ways: through convective clusters that form along the ITCZ and which, as they spread westwards, reach the Amazon basin, and through the interaction of the trade winds with the circulation of the WB, forming ILs that enter the continent and regenerate during their westward propagation, many of which reach the Andes (KOUSKY, 1980; COHEN et al 1989; COHEN et al., 1995).

In studies by Munnich and Neelin (2005), using monthly data based on satellite precipitation, sea surface temperature (SST) and sea level height, they found that there is a correlation of precipitation anomalies over equatorial South America in May with simultaneous SST anomalies in the equatorial Atlantic and Pacific, which would support a causal chain in which ENOS induces wind stress anomalies in the western equatorial Atlantic and affects Atlantic equatorial ocean dynamics. These correlations show strong seasonality, apparently appearing within the atmospheric links of the chain. This path and the influence of the equatorial Atlantic ATSM on South American precipitation in May, which appear to be independent of those of the northern tropical Atlantic, show that the NEB is affected by the tropical North Atlantic.

The ITCZ represents the atmospheric system with the potential to cause large amounts of rainfall over tropical continental areas, including large metropolises where the population is highest. Among the largest cities located in the region where the ITCZ acts, the city of Belém do Pará stands out. In this region, the month of March is climatologically known as the wettest and the ITCZ, which is a planetary-scale atmospheric system, is in its southernmost position (around 02°S on average) and is responsible for numerous extreme precipitation events (FERREIRA et al., 2013).

Ferreira et al., 2015, when studying the spatial and temporal variability of precipitation in the eastern Amazon during the rainy season, using Empirical Orthogonal Functions (EOF) employed in Long Wave Radiation (LWR) anomalies, showed that the ITCZ is the main mode of climate variability over the Intertropical Atlantic Ocean and is the main precipitation-inducing system in the eastern Amazon during the rainy season. The study revealed a dynamic pattern related to the ITCZ in the South Equatorial Atlantic, with the presence of the South Adantic SST gradient, which in turn contributes directly to abundant rainfall in the rainy season in eastern Amazonia, so that positive rainfall anomalies are observed especially in

Maranhão and north-eastern Pará.

These significant rainfall events generally directly affect the local population, causing a pronounced rise in Amazonian rivers, flooding in several municipalities, among other conditions.

Recently, Freitas et al. (2017) investigated the extreme shifts of the ITCZ in the Southern Maritime Continent (SMC) region in austral spring during the period 1979-2015 and the results showed that during the extreme shifts of the ITCZ to the south, the Walker circulation is stronger, leading to the strengthening of the north-westerly equatorial winds in the Indian Ocean, which pumps humid air and stimulates diabatic heating in the SMC due to the positive local TSM anomalies present in this region. Conversely, during extreme northward shifts, the Walker circulation would become weaker, not allowing the north-westerly equatorial winds in the Indian Ocean to fulfil their role, which associated with the local negative ATSM, leads to drier conditions in the CMS region.

Another interesting result found by the authors was that a strong meridional or zonal index does not necessarily imply a greater shift. The combination of ENOS events and a very strong Indian Ocean dipole can lead to larger shifts of the ITCZ further south. In addition, the southernmost ITCZ can act as a bridge between the Indian Ocean dipole and ENOS.

2.1.8. Large-scale Climatic Mechanisms that Influence the Intensity and Position of the ITCZ

2.1.8.1. El Nino-Southern Oscillation

The main studies on ENOS episodes have been based on the ATSMs that occur in the Eastern Equatorial Pacific, over the Nino 3 or Nino 3.4 region. These episodes significantly alter atmospheric circulation, precipitation (KOUSKY et al., 1984; ROPELEWSKI and HALPERT, 1987, 1989; ACEITUNO, 1988; GRIMM et al., 2000; GRIMM, 2003, 2004), extreme precipitation events (LIEBMANN et al., 2001; CARVALHO et al., 2002; GRIMM and TEDESCHI, 2009; ROBLEDO et al., 2013), surface temperature (BARROS et al., 2002), river flow (CAMILLONI and BARROS, 2000; BERRI et al., 2002) and agriculture and hydroelectric production (COLLISCHONN et al., 2001; BERBERY and BARROS, 2002; TUCCI et al., 2003) in South America.

The spatial and temporal variability of rainfall over the Amazon during the rainy season is directly influenced by the large-scale climate mechanisms that occur in the two adjacent tropical oceans: the Equatorial Pacific Ocean and the Tropical Atlantic Ocean during the autumn months. These modes trigger interactions between the ocean and the atmosphere and are basically related to: the ENOS Cycle over the Pacific Ocean and the Dipole Pattern (HASTENRATH and HELLER, 1977; MOURA and SHUKLA, 1981; NOBRE and SHUKLA, 1996; DE SOUZA and NOBRE, 1998) or the Meridional Gradient Phases of the SST anomalies in the intertropical basin of the Atlantic Ocean (SOUZA and ROBINSON, 2004; SOUZA, 2005).

The ENOS cycle has two well-known phases, the EI Nino (EN) and La Nina (LN), in which EN conditions are associated with large-scale anomalous warming in the ocean waters over the central-eastern Equatorial Pacific, extending for at least five months between summer and autumn.

Conversely, LN conditions are related to the anomalous cooling of the equatorial ocean waters over the central-eastern basin. In EN years, the descending branch of the Walker cell moves into the region over the Amazon, inhibiting convection during what would be the rainy season (January, February and March) in the northern region, more precisely in central Amazonia. The north-easterly trade winds are much weaker, thus reducing the flow of moisture from the oceans into the Amazon region (PHILANDER, 1990; TRENBERTH, 1997; DRUMOND and AMBRIZZI, 2005; KAYANO et al., 2005; KAYANO and ANDREOLI, 2006).

2.1.8.2. Atlantic Dipole Pattern

The Dipole Pattern has two opposing phases, called the Positive Dipole phase and the Negative Dipole phase. The positive Dipole phase is characterised by the simultaneous presence of positive-negative SST anomalies over the north/south basin of the tropical Atlantic Ocean. The negative dipole phase has the opposite configuration.

Souza et al. (2000) assessed the influence of the Dipole Pattern on precipitation in the Amazon, concluding that in years with EN patterns in the Pacific and a positive dipole in the tropical Atlantic, there is a stronger effect on reducing precipitation over the centre-north and west of the Amazon than when there is only the occurrence of an EN.

For Hastenrath (2002), there is no "seesaw" of the SST in the Atlantic Ocean. The main thing is the thermal gradient of the TSM between the hemispheres, such that in years with a stronger positive meridional gradient in the south-north direction, the ITCZ is positioned further north, resulting in a year of below-normal rainfall in the NNEB; when very active, the ITCZ increases the pressure gradient between the equator and the subtropics (MOLION and BENARDO, 2002).

Sousa et al. (2002) observed a strong correlation between rainfall in Pará and the EN and LN phenomena. Notably, in LN years with a negative dipole, rainfall in Pará is favoured. The EN with a positive dipole is associated with reduced rainfall (drought). When the south and north Atlantic are neutral, but with abnormally cold waters, there is a reduction in rainfall and the opposite when the waters are abnormally warm.

Ferreira et al. (2015) found that Amazon climate variability is influenced by large-scale tropical climate modes associated with ENOS in the Pacific Ocean and SST in the Atlantic Ocean, and that the Atlantic gradient appears dominant during austral autumn. Typically, a northward (southward) gradient presents simultaneous positive/negative (negative/positive) SST anomalies in the north/south tropical basin of the Atlantic Ocean. These meridional gradients would hydrostatically control sea level pressure, wind patterns and moisture convergence at lower levels, influencing the latitudinal positioning of the ITCZ during its seasonal migration to the South Atlantic.

2.1.8.3. Madden-Julian Oscillation or 30-60 Oscillation

Another large-scale dynamic mechanism that directly influences rainfall patterns in the Amazon is the Madden-Julian Oscillation (OMJ) (MADDEN; JULIAN, 1994). This mechanism acts on an intraseasonal scale and is a low-frequency phenomenon dominant in most of the tropical troposphere, characterised by an equatorial atmospheric wave propagating eastwards along the equator in a complete cycle around the globe for approximately 30-60 days (MADDEN and JULIAN, 1994), directly influencing rainfall in the eastern Amazon and NEB (SOUZA and AMBRIZZI, 2006).

The variability in the position and cloudiness of the ITCZ may also be due to the OMJ (REPELLI et al., 1998), which can intensify the convective activity of the ITCZ during the passage of this oscillation's trough and weaken it during the passage of the Christian one.

The NOAA (National Oceanic Atmospheric Administration) is currently monitoring the development of the OMJ, as it has already been recognised as one of the main weather modulators in the equatorial region and can interfere with the weather systems that act on it (ZHANG, 2005).

Souza and Ambrizzi (2006) studied the modulation of precipitation in Brazil and observed that this intra-seasonal oscillation is responsible for the increase in precipitation over the Amazon, interacting with the systems responsible for sub-monthly variations, such as the SACZ and ITCZ (SOUZA et al., 2005).

Vitorino (2003) and Vitorino et al. (2006) showed by applying the Morlet wavelet transform (WT) to the pressure fields at mean sea level, emerging ROL and relative vorticity at high levels, that this oscillation influences each sub-region of Brazil differently.

Moraes and Sá (2006) applied TO to a 10-year time series and observed that the intraseasonal scale was the most energetic in this period, with the OMJ being the main atmospheric mechanism responsible for this energy. Another point highlighted by the authors was the frequency between five (5) and fifteen (15) days. During the same periods of atmospheric modulation by the OMJ, these time scales, related to the SACZ, also showed signs of high intensity, reaffirming what was written earlier about the link between these two meteorological phenomena.

The OMJ is not only related to systems on scales smaller than its own. In addition to its interaction with the SACZ and ITCZ, it is connected to another important mechanism for modulating weather and climate in the Amazon region, ENOS. During its positive ENOS phase, the eastern equatorial Pacific tends to become anomalously warm, displacing the convective cell from Oceania to the east. As a result, the OMJ weakens to the point of disappearing. On the other hand, during the negative phase (LN) the waters over Oceania become warmer, convection is intensified, favouring the generation of OMJ (LAU and CHAN, 1988; PHILANDER, 1990; VITORINO et al., 2006).

Oliveira (2012) studied the possible modulation effects of the OMJ on various meteorological variables using ROL and zonal wind component reanalysis data and concluded that the occurrence of the OMJ in the rainy season in the eastern Amazon could contribute to the intensification of convection and precipitation in the region with possible adverse consequences for society, such as the generation of disastrous climatic situations, increased precipitation, intensified winds, increased possibility of flooding and other natural disasters that affect the economy, transport and the well-being of the population.

Oliveira et al. (2015) investigated the influence of the OMJ on local climatic characteristics in the eastern Amazon using six variables. Maximum temperature, average ROL temperature and insolation showed the most significant difference between the two phases of the OMJ (all above the 95 per cent level). These results indicate that the OMJ is capable of modifying the dynamics of the local atmosphere. The increase in convective activity during its active phase is characterised by an increase in the amount of clouds, reducing the daytime temperature and generating fewer hours of sunlight at the surface. The opposite occurs during the OMJ's inactive phase. However, precipitation is a more complex variable because various meteorological systems and mechanisms can influence its daily accumulated value. The small difference between the average precipitation values in each of the OMJ phases may be a response to these various, and sometimes conflicting, influences.

Shimizu and Ambrizzi (2016) investigated the combined response of the phases of the two distinct phenomena, ENOS and OMJ, over South America to obtain the average temperature and precipitation patterns for the months of November to March (austral summer) and May to September (austral winter). The results showed that the patterns of precipitation and temperature anomalies observed during ENOS phases, without the concomitant occurrence of the OMJ, can be strengthened or weakened during events where ENOS and OMJ occur simultaneously. Furthermore, the effect on the anomaly patterns in these events depends on the phase of the JMO.

2.2. Definition of the Method Used to Map and Spatialise Rainfall in Pará.

- Cluster Analysis

This term was first used by Tryon (1939) and the technique is in fact a set of classification processes (algorithms) that aim to group data according to their similarities.

There are many different methods that can be used to carry out a Cluster analysis, these methods can be classified as: Hierarchical and non-hierarchical methods (CORNISH, 2007).

- Hierarchical methods

- Agglomerative methods, in which individuals start in their own separate cluster. The two "closest" (most similar) clusters are then combined and this is done repeatedly until all the individuals are in one cluster. In the end, the optimal number of clusters is then chosen from all the cluster solutions.

- Divisive methods, in which all subjects start in the same cluster and the above strategy is applied in reverse until each subject is in a separate cluster. Agglomerative methods are used more often than divisive methods.

Hierarchical methods generate strictly nested structures and are often presented graphically using a dendrogram (GORE, 2000), which is the most usual way of representing the results of hierarchical algorithms and intuitively shows the order of the grouping (LINDEN, 2009).

Four different agglomerative methods are frequently found in the literature.

These methods differ mainly in how the elements are compared to each other in each successive phase of Clustering. Sneath (1957) and McQuitty (1987) proposed the single linkage procedure, in which an object is included in a cluster that shares a sufficient degree of similarity with at least one member of a cluster. This method has also been referred to as primary linkage, the method of minima, and nearest neighbour cluster analysis (JOHNSON, 1967; LANCE and WILLIAMS, 1967).

In this method, the distance between two clusters is defined as the distance between the two closest members, or neighbours. This method is relatively simple, but is often criticised because it doesn't take into account the cluster structure and can result in a problem called chaining whereby the groups end up being long and unkempt. However, it is better than the other methods when the natural clusters are not spherical or elliptical in shape.

A second agglomeration method, which refers to the most complete linkage analysis, is the maximum method, or most distant neighbour analysis (HORN, 1943), which is the opposite of the single linkage method in that the distance between poles is defined as the distance between the most distant pair of cluster members. In this case, the distance between two clusters is defined as the maximum distance between the members - that is, the distance between the two individuals who are furthest apart. This method tends to produce compact clusters of similar size, but, like the nearest neighbour method, does not take into account the cluster structure. It is also very sensitive to discrepant observations.

The average linkage method (SOKAL and MICHENER, 1958) joins elements to existing clusters based on the average degree of similarity between the element and the existing members of the cluster. The average linkage method actually represents a set of related methods that differ in terms of how the "averaging" is conducted, i.e. the distance between two clusters is calculated as the average distance between all pairs of individuals in the two groups. This is considered to be a fairly robust method. Romesburg (1984) describes the unweighted pair-group method which uses arithmetic averages of pairs to recalculate the distance matrix.

Somewhat related to the unweighted pair-group method is the centroid method of cluster analysis. In this method, distance is defined as the distance between the centres of the groups. Since the arithmetic mean process recalculates distance matrices by averaging previously determined distance values, the recalculation of a distance matrix in the centroid method is preceded by updates to the raw data matrix.

Here, the centroid (average value for each variable) of each cluster is calculated and the distance between the centroids is used. Clusters whose centroids are closer together are merged. This method is also relatively robust.

A final group of agglomerative methods frequently used by researchers are sum-of-squares or minimum variance methods. Ward (1963) is probably the most widely used minimum variance method, although other sum-of-squares procedures were available to the researcher (FRIEDMAN and RUBIN, 1967; WISHART, 1969). The relative proximity of a set of objects can be described using the concept of the sum of squares (squaring the sum of the distances of each object from the mean value of the cluster). Using Ward's method, the cluster results in the smallest increase in the sum of squares being formed during each step. All possible combinations of fragmentation formations are considered in each subsequent step.

- Non-hierarchical methods (often known as k-means clustering methods).

In these methods the desired number of groups is specified in advance and the 'best' solution is chosen.

Non-hierarchical cluster analysis tends to be used when large data sets are involved. It is sometimes favoured because it allows the individual to move from one set to another (this is not possible in hierarchical cluster analysis where a subject, once assigned, cannot move to a different group). Two disadvantages of non-hierarchical cluster analysis are: (1) it is often difficult to know how many groups you are likely to have and therefore the analysis may have to be repeated several times and (2) it can be very sensitive to the choice of initial cluster centres. Again, it may be worth trying different ones to see what impact this has.

One possible strategy to adopt is to use a hierarchical approach initially to determine how many groups there are in the data and then use the group centres obtained from these cluster centres as initials in the non-hierarchical method (CORNISH, 2007).

Filho et al. (2005), using *cluster* analysis, identified 25 homogeneous zones (regions) in Brazil in terms of the distribution of rainfall probability. Their work was developed with the aim of supporting future studies aimed at climate risks in the organisation of the national agricultural space, especially in areas with regular rainfall data. In carrying out this diagnosis, the Clustering method was also used to regionalise rainfall data.

In a study carried out in the state of Minas Gerais, Torres (2007) points out that the homogenisation of rainfall information through cluster analysis aimed at regionalising this data is one of the most successful statistical applications for making up for the lack of rainfall data in a river basin. This idea is shared by Tucci (1993) who emphasises that through hydrographic regionalisation using statistical methods, it is possible to arrive at homogeneous groups with similarities in the hydrological response of each region. According to this author, regionalisation is used to identify areas with similar behaviour, so that hydrological responses can be correlated, thus facilitating the transfer of information between regions. This favours the correction of flaws in existing observations.

Melo Neto (2007a), when analysing the processes that have contributed to soil erosion on one of the slopes of the River Jaguaribe in the municipality of Conceição do Almeida in the state of Bahia, regionalised the rainfall data from seven meteorological stations in order to obtain information in his future analyses that could be transposed to the stations with incomplete data. In order to achieve its objective, a series of statistical techniques were used, in particular Ward's Cluster Analysis.

In another published study, Melo Neto (2007b) used rainfall regionalisation to analyse the main physical factors influencing the different amounts of rainfall that hit his study area between 1970 and 2006.

According to the author, this procedure was important for diagnosing the role of relief in rainfall distribution.

Neto and Villa Gois (2007) also used cluster analysis to identify regions where rainfall is similar to each other, i.e. homogeneous regions in the Da Dona river basin. Their aim with the homogeneous regions is to verify the role of topography in the distribution of rainfall in this basin and, from there, to diagnose the environmental impacts caused by the sediment carried by rainfall to the dams built in this catchment area.

Santos et al. (2008), using Ward's clustering method to map homogeneous rainfall in the Goiana-PE river basin, verified the main physical factors, internal to the basin, that influence the spatialisation of rainfall. His results were, after finding three homogeneous regions, that the region is strongly influenced by maritime and topographical factors.

CHAPTER 3

MATERIALS and METHODS

3.1. Study Area

The study area covers the state of Pará (northern Brazil) (Figure la) and specifically its capital Belém. Pará is the second largest state in Brazil with an extension of 1,248,042.515 km^2 , divided into 144 municipalities, and is bordered by Suriname and Amapá to the north, the Atlantic Ocean to the north-east, Maranhão to the east, Tocantins to the south-east, Mato Grosso to the south, Amazonas to the west and Roraima and Guyana to the north-west. The main rivers are the Amazon, Tapajós, Tocantins, Xingu, Jari and Pará (Figure 1c). The city of Belém, bathed by the Guamá River and the Guajará Bay (captured by satellite - Figure 1b), is the rainiest capital in Brazil (SOMAR, 2015), has an area of 1,064.918 km^2 , an altitude of ten metres above sea level, thirty-nine islands in its territory, situated in the Atlantic Ocean and is bordered by the municipalities of Ananindeua, Marituba, Santa Barbara do Pará and Barcarena, as well as the Marajó Bay.

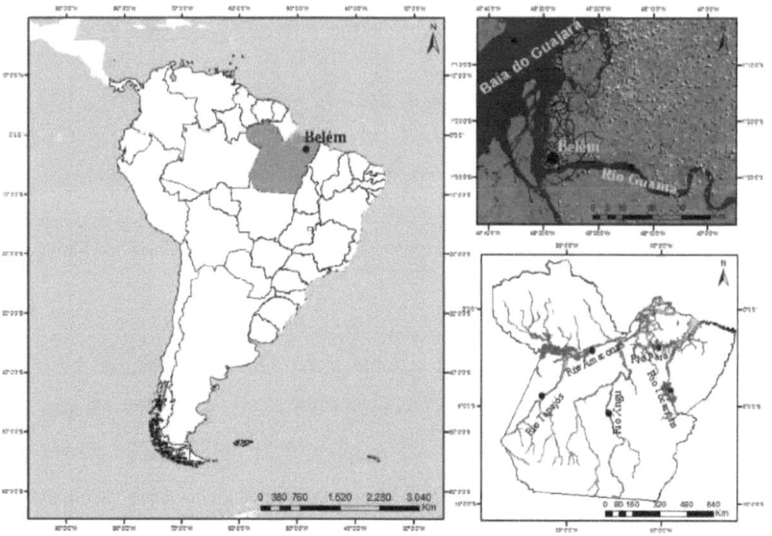

Figure 1: (a) Map of Brazil, highlighting the state of Pará in green, (b) Satellite image of the city of Belém-Pará, the banks of the Guamá River and Guajará Bay. (c) Map of the state of Pará and its main rivers and tributaries. Source: IBGE, CODEM and ANA.

The climate in Belém is directly influenced by the Amazon rainforest, its rivers and its proximity to the Atlantic Ocean. As a region where rainfall is constant, the city has two seasons: the rainy season (December to May) and the less rainy season (June to November) (Figure 3), as described by Nechet (1984), Figueroa and Nobre (1990), Bastos et aL, (2002) and Oliveira et al., (2003).

In general, the rainy season begins in December and lasts five to six months, with the wettest month being March, which is explained by the intensification of local effects associated with the presence of the ITCZ, causing the wettest quarters of March, April and May and December, January and February. The less rainy season, where convective rains in the form of isolated rain showers predominate, covers the remaining months of the year, with the period of lowest climatological monthly rainfall occurring between August and November (ALBUQUERQUE et al., 2010) (Figure 2).

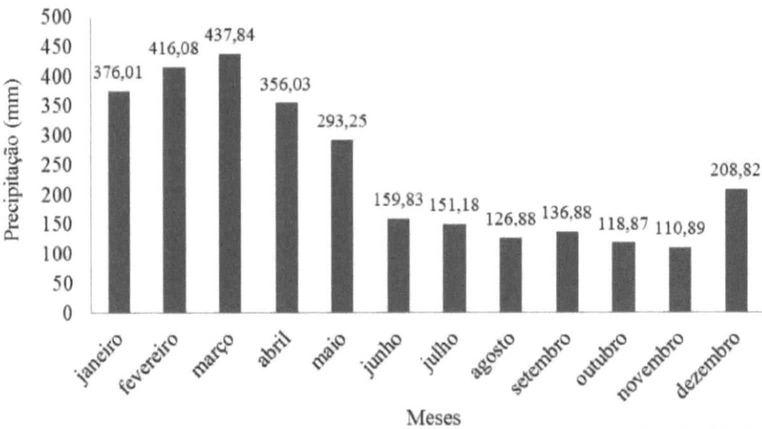

Figure 2: Climatological averages of monthly rainfall (1961-1990 average) representative of Belém do Pará for the months of January to December, adapted by the author. Data source: INMET

The spatial and temporal distribution of rainfall in the Amazon region has been the subject of several studies, such as Figueroa and Nobre (1990); Marengo (1995); Souza and Ambrizzi (2003); Lopes (2009); Amanajás and Braga (2012), which have shown the importance of rainfall in the region, as it can influence the behaviour of other atmospheric variables, such as temperature, relative humidity, wind, among others (FISCH and MARENGO, 1998).

3.2. Data

This work uses two sets of data. One set of hourly rainfall data from five (5) INMET automatic stations, and another set of daily data from twenty-five (25) stations distributed throughout the state of Pará, fifteen (15) of which are INMET stations and ten (10) from the National Water Agency (ANA). For the periods 2003 to 2015 and 2008 to 2015. Figure 3 illustrates the map of the state of Pará with the numbering of each station and, below, the list with the name of each of the stations and their respective coordinates (Table 1).

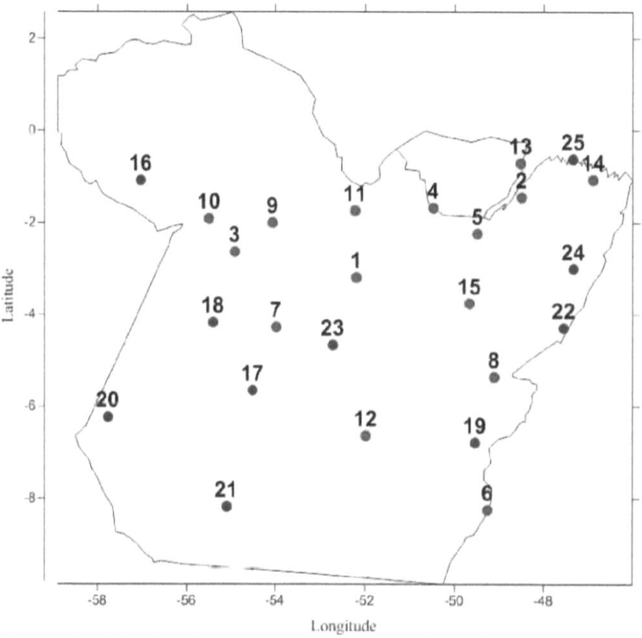

Figure 3: Spatialisation of rainfall stations in the state of Pará. Balls in blue represent ANA stations and balls in red represent INMET stations.

Table 1. Representation of the selected stations and their respective coordinates.

Number	Station	Municipality	Latitude	Longitude
1	ALTAMIRA	ALTAMIRA	-3,2	-52,2
2	BELEM	BELEM	-1,45	-48,5
3	BELTERRA	BELTERRA	-2,63	-54,93
4	BRIEF	BRIEF	-1,68	-50,48
5	CAMETA	CAMETA	-2,24	-49,49
6	AARAGUAIA CONCEPTION	AARAGUAIA CONCEPTION	-8,25	-49,26
7	ITAITUBA	ITAITUBA	-4,27	-53,98
8	MARABA	MARABA	-5,36	-49,11
9	MONTEALEGRE	MONTEALEGRE	-2	-54,06
10	OBIDOS	OBIDOS	-1,91	-55,51
11	PORTOMOZ	PORTOMOZ	-1,74	-52,23
12	SÃOFELIXXINGU	SÃOFELIXXINGU	-6,64	-51,99
13	SOURE	SOURE	-0,71	-48,52

14	TRACUATEUA	TRACUATEUA	-1,07	-46,9
15	TUCURUI	TUCURUI	-3,76	-49,67
16	PORTEIRA WATERFALL	ORIXIMINÁ	-1,08	-57,04
17	CAJUEIRO	ALTAMIRA	-5,65	-54,52
18	CUPARI	RURÓPOLIS	-4,17	-55,42
19	SANTA ELISA FARM	SAPUCAIA	-6,79	-49,54
20	JACAREACANGA	JACAREACANGA	-6,23	-57,77
21	KM947BR163	ITAITUBA	-8,18	-55,11
22	KM ZERO PA 70	DOM ELISEU	-4,29	-47,56
23	MONTE ALEGRE DO XINGU	ALTAMIRA	-4,66	-52,72
24	PARAGOMINAS	PARAGOMINAS	-3,01	-47,34
25	SALINOPOLIS	SALINOPOLIS	-0,62	-47,35

3.3. Data Description

Initially, hourly rainfall time series were used for the Belém station, provided by INMET through the 2nd Meteorological District (2° DISME). We then chose daily rainfall series from INMET stations, but INMET only has 15 stations throughout Pará, so we opted to select ANA stations to cover the entire state. Two periods were chosen. The first was from 2003 to 2015, with hourly rainfall series from the Belém station, provided by the 2° DISME. These years were chosen because the station was founded in 2003. The observed accumulations, for example, are accumulated from 12h00 UTC (09h00 HL) to 12h59 UTC (09h59 HL), which would correspond to the time of 13h00 UTC (lOhOO HL) and so on.

The times used were from OOhOO to 23h00 UTC. To do this, the percentage contribution of rainfall in relation to the monthly total in Belém was analysed using descriptive statistics, with the aim of summarising various values of the same nature, so as to have a global view of the variation in values (GUEDES et al., 2010) and drawing up histograms. For this purpose, spreadsheets were used from the @ Microsoft Office Excel 2010 software.

The second set of data used was daily data from the years 2008 to 2015, which is accumulated rainfall over 24 hours, observations made all day at 12:00 UTC (09:00 HL), for regionalisation (determining homogeneous regions). This was also chosen because the daily data from the other stations, provided by INMET, were founded after 2008. The aim of the cluster analysis was to carry out a study of the state of Pará, delimiting the regions with similar rainfall patterns. Ward's hierarchical agglomerative method (WARD, 1963) was used to carry out the regionalisation, with Euclidean distance as the measure of dissimilarity. This is one of the most widely used methods in studies to minimise the variability between data, grouping together elements that should be homogeneous and the elements of other different groups that should be highly heterogeneous (HAIR et al., 2005).

In order to compare the phases of ENOS with the interannual variability of precipitation, the Nino Oceanic Index (INO) of the *Climate Prediction Centre of the National Oceanic and Atmospheric Administratation* (NOAA) was used, available on the NOAA website. This index defines the phases based on the ATSM as weak (ATSM of 0.5oC to 0.9oC), moderate (1.0oC to 1.4oC) and strong (>1.5oC).

Tables (attached) of rainfall frequency by hour, for each month and for the period, were constructed using the available data. From these, histograms of the number of days with rain according to time were drawn up and, using these, frequency polygons were drawn up representing the number of days with rain according to time, for each month and for the period considered. This made it possible to ascertain the times of day with the highest and lowest frequency of rainfall in each month and over the period in question. A frequency polygon was then generated for the first half of the year, when most of the rainfall (including the greatest volume of rain) is concentrated over the region, mainly in the months of March and April (KOUKY, 1979; HASTENRATH and HELLER, 1977; STRANG, 1972), which is therefore considered the "rainy" half of the year, and another for the second half of the year (considered less rainy, or "dry").

The graphical form used to represent the final result of the various groupings is the dendrogram. It shows lines connected according to levels of similarity that group together pairs of individuals or variables. The dendrogram shows graphically how the clusters are combined at each step of the procedure until they are all contained in a single cluster (LANDIM, 2011).

The analysis procedures, such as the dendrogram representation of the results, were carried out using SPSS 13.0 (Statistical Package for Soil Sciences) software.

With the selection of this data, centimetres were calculated just to group the stations and to map homogeneous rainfall. To do this, the data was processed and organised to calculate the daily averages. This spreadsheet was used to perform Ward's hierarchical clustering method and the Surfer 8.0 programme was used to make the maps. Subsequently, some rainfall stations with rainfall similar to those in the city of Belém were selected to analyse the stations with the same rainfall regime in the afternoon using hourly data.

3.4. Methodology

3.4.1. Descriptive Statistics

Using descriptive statistics, it is possible to obtain essential characteristics for forming relative frequency histograms of a sample of hydrological data (NAGHETTINI and PINTO, 2007). Measures of central tendency were calculated for this study.

- Measure of Central Tendency

The mean is the most frequently used measure of position and has an important theoretical significance in estimating samples. It is calculated using the following formula:

$$\acute{X} = \frac{x_1 \ldots\ldots\ldots\ldots, x_n}{N} = \frac{1}{N}\sum_{1=0}^{n} x_i \qquad (1)$$

Where X is the mean; N is the sample size and Xi to x_n are the sums of the sample.

3.4.2 Cluster methods

There are two types of group classification methods or algorithms. One is the hierarchical method, where the groups are partitioned from a minimum of initially undefined groups, the majority groups are divided into minority subgroups grouping together those individuals with similar characteristics. The final structure of the classes is presented in the form of a classification tree (dendrogram) which presents an objective summary of the results (EVERITT, 1993; WILKS, 2006).

- Ward's method

This is a hierarchical method that uses the Euclidean distance to measure the similarity or dissimilarity between individuals, i.e. the distance between individuals X and Xj is given by (EVERITT, 1993):

$$d_{ij}=|x_i-x_j|=\left[\sum_{k=1}^{p}(x_{i,k}-x_{j,k})^2\right]^{\frac{1}{2}} \qquad (2)$$

The Ward method forms groups by minimising dissimilarity, or minimising the total sum of squares within groups, also known as the sum of squares of deviations (SQD). At each stage of the procedure, groups are formed in such a way that the resulting solution has the lowest SQD within groups. In these stages, the unions of all possible pairs of groups are considered, and the two that result in the smallest increase in SQD are grouped together until all the groups form a single one, bringing together all the individuals (EVERITT and DUNN, 1991). This method also proposes that at any stage of the analysis, the loss of information resulting from the grouping of elements between groups is measured by the SQD of each point to the average of the group to which it belongs (EVERITT, 1993).

$$SQD=\sum_{i}^{n}x_i^2-\frac{1}{N}\left(\sum_{i=1}^{n}x_i\right)^2 \qquad (3)$$

Where *n* is the total number of elements in the cluster and Xi is the i-th element in the cluster.

Average monthly rainfall values were organised in the form of a Pnxp matrix, where the Py element represented the value of the i-th variable (location) of the j-th individual (day). Thus, each row vector represented a location and each column vector a day.

After grouping the days with similar rainfall, Ward's method was applied once again to determine, within each of the daily periods defined, the homogeneous regions of locations with similar rainfall. In this case, the Pnxp matrix was organised in such a way that the individuals represented the locations and the grouped days (periods) represented the variables.

CHAPTER 4

RESULTS and DISCUSSION

4.1. Analysing Precipitation in the State of Pará

4.1.1. Variability of Average Annual Precipitation Totals in the State of Pará

The state of Pará is part of the Amazon region, with a large territorial dimension. The spatio-temporal variability of rainfall in the region is influenced by local effects, meso and large scale systems such as convective clusters, LI, transient meteorological systems (quasi-stationary troughs), ITCZ and SACZ (COHEN et al., 1989; QUADRO, 1994; COELHO et ah, 2004). Thus, the action of these different meteorological systems on different time scales results in a high but inhomogeneous rainfall distribution. Figure 4 shows the spatial variability of the average annual rainfall totals in Pará during the period studied. The highest volumes of more than 2,000mm/year are found in the northeast and northwest of the state, and the lowest volumes of less than 1,700mm/year in the centre-south of the state.

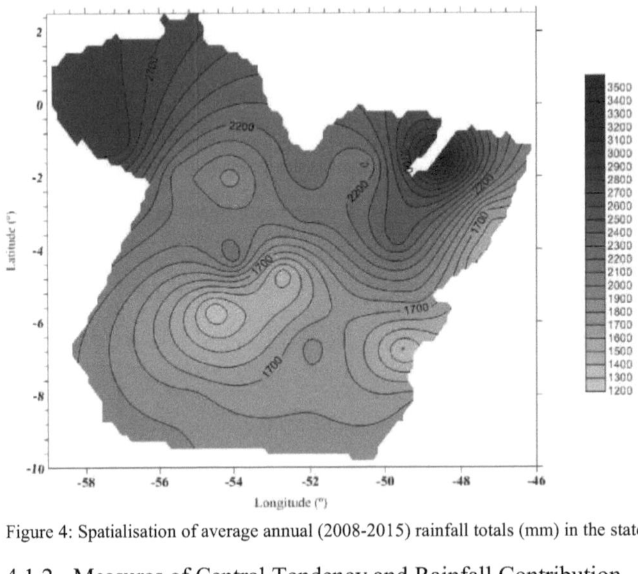

Figure 4: Spatialisation of average annual (2008-2015) rainfall totals (mm) in the state of Pará.

4.1.2. Measures of Central Tendency and Rainfall Contribution

A climatic approach to rainfall was carried out in order to calculate the arithmetic mean, providing a quantitative view of the percentage contributions of rainfall in the years 2003 to 2015. Figure 5 shows the behaviour of the hourly rainfall cycle in the city of Belém-PA and the average percentage contribution of the monthly total from 2003 to 2015.

Given that Belém has well-defined characteristics, such as surface heating throughout the day and a greater amount of precipitation in the late afternoon and early evening, the figure shows greater contributions between 3pm and 6pm HL (average of 14%). What stands out in the figure is that between the end of the night, during the early hours and in the early morning (from 10pm to 12am HL) the average contribution is minimal, reaching values below 2%. In the early afternoon (13h00 HL) this contribution gradually increases and at 16h00 HL it reaches its maximum peak (17%). At this time there are favourable conditions for the development of cumuliform clouds (cumulus and Cbs, i.e. with a strong vertical dimension), which then form heavy precipitation mainly in the afternoon, thus increasing the contribution, and from 17h00 HL onwards, it can be seen that these values decrease.

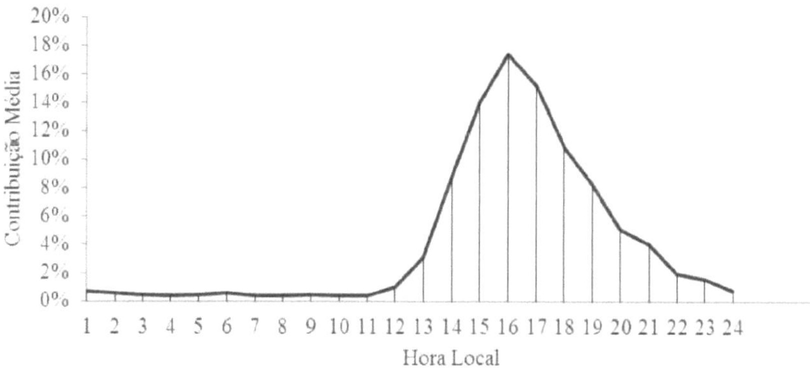

Figure 5: Average percentage rainfall of the daily contribution for the years 2003 to 2015.

4.1.3. Analysing the Percentage Contribution of the Rainiest Months.

Figure 6 shows the hourly variation in the average percentage of the rainfall contribution to the monthly total for Belém-PA, from 2003 to 2015, referring to the wettest months.

The figure shows that in all the months the greatest contributions were in the afternoon, with values above 10% and a maximum peak around 16:00 HL. This maximum can be explained by the action of the ITCZ and mesoscale effects, such as the LI that propagate westwards as a line of Cbs and associated with the BM, formed in the afternoon, cause intense rainfall, more frequently between April and August, and also the formation of MLF, which occur between 12:00 and 14:30 HL, and between 14:30 and 20:00 HL, the formation of micro lines of maritime instability.

The observed characteristics follow the same pattern in all months, with values below 2 per cent from dawn to mid-afternoon, with peaks in the late afternoon and early evening. As a result, the expectation of precipitation in the morning is very low. On the other hand, from 3pm HL onwards, the surface heating and breeze circulation on the coast leaves the atmosphere favourable for the formation of clouds with high tops, increasing the likelihood of intense and rapid precipitation.

In December, due to the rains returning to the afternoon, this maximum peak is higher (26%) and in May it is lower, corresponding to 14%. There is a small increase in the contributions in April and May, in the evening period compared to the other months, which in the figure is below 2%. This may be related to the shift of the ITCZ to the HN in these months.

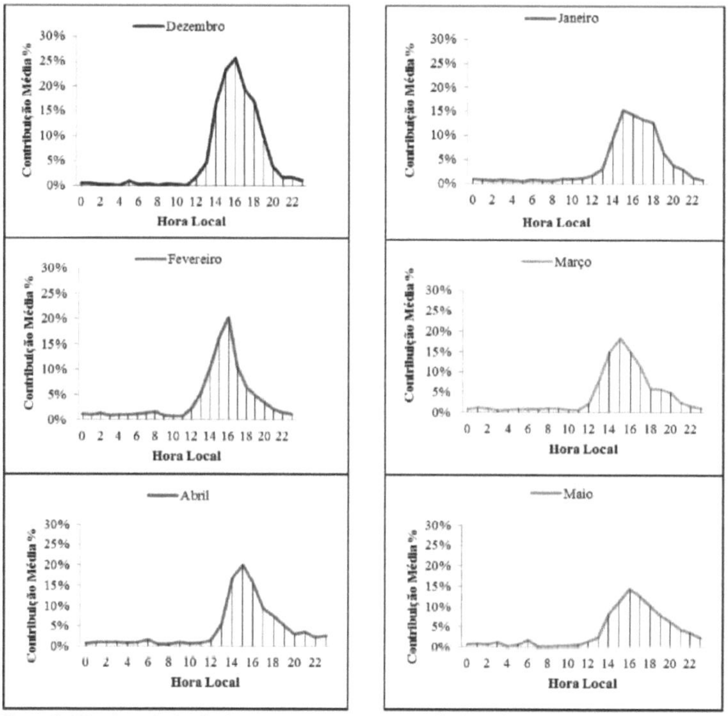

Figure 6 - Hourly variation in the average percentage contribution of total monthly rainfall during the wettest period (December to May) in Belém.

4.1.4Analysing the Percentage Contribution of the Least Rainy Months

Figure 7 shows the hourly cycle of the average percentage contribution of total rainfall in the least rainy months in Belém, from 2003 to 2015.

It can be seen that from September to November, considered to be the dry season, precipitation is generally caused by mesoscale phenomena (BASTOS et al., 2002), so the greatest contributions are in the afternoon, corresponding to 26% of the rainfall total and the contributions at night reach values below 2%, similar to figure 6. The hourly cycle shows two peaks, the first in September and the second in November with an average percentage of 25% of the rainfall total, between 16:00 and 17:00. In the other months (June, July, August and October) this contribution was around 15%.

In the months of June, July and August, there was a slightly higher contribution at night, around 5%, which is probably related to SCMs, which are systems that originate at night, as well as MLFs, which are frequent in August, and the highest number of LI occurrences seen between April and August, during which time LIP formation stands out (COHEN el al., 1989).In addition to the fact that this curve shows contributions at night in the months of June, July and August, which are the end of the rainy season, they may also have been caused by local effects such as BT and BM and by DOL, since these systems are most frequent in the afternoon and consequently the largest contributions are in this period, and so the smallest contribution at night is seen in the months of October, November and December, when the ITCZ is further north of its climatology.

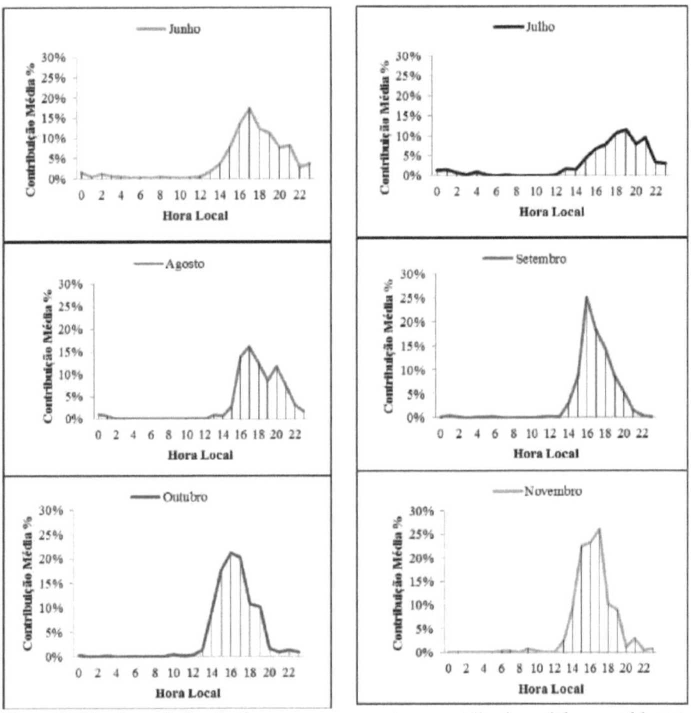

Figure 7: Hourly variation in the average percentage contribution of the monthly total for the less rainy period (June to November) in Belém.

4.1.5. Interannual and Monthly Precipitation Analyses for the Years 2003 to 2015 and ENOS Events in the Precipitation of Belém-PA.

Figure 8 shows the inter-annual variability of accumulated rainfall between 2003 and 2015. This accumulation ranged from 2500 mm/year to 3800 mm/year. The years with the highest rainfall accumulations were 2006 with 3663.8 mm, 2013 with 3775.6 mm, 2011 with

3592.2 mm and 2012 with 3563 mm of rain. With the exception of 2006 and 2013, which were neutral years, 2011 and 2012 were under the influence of the LN phenomenon, which increases rainfall in the city of Belém. The lowest volumes of rainfall occurred in 2003 with 2769.4 mm of rain and 2004 with 2940.2 mm. These years were under the influence of the EN phenomenon, which inhibits rainfall in Belém-PA. These results corroborate those of CPC/NOAA (2010), who observed that ENOS was negative in 2003, 2004-2005 and 2006-2007 (beginning). And in the years 2007(end)-2008 the index was positive, configuring an LN event, as shown in Table 2.

Table 2: Occurrence of EI Nino and La Nina events between 2003 and 2015.

Years	DJF	JFM	FM A	MAM	AMJ	MJJ	JJA	JAS	ASO	SON	OND	NDJ
2003	0.9	0.7	0.4	0	-0.2	-0.1	0.1	0.2	0.2	0.3	0.3	0.3
2004	0.3	0.3	0.2	0.1	0.2	0.3	0.5	0.6	0.7	0.7	0.6	0.7
2005	0.7	0.6	0.5	0.5	0.3	0.2	0	-0.1	0	-0.2	-0.5	-0.7
2006	-0.7	-0.6	-0.4	-0.2	0.0	0.0	0.1	0.3	0.5	0.7	0.9	0.9
2007	0.7	0.4	0.1	-0.1	-0.2	-0.3	-0.4	-0.6	-0.9	-1.1	-1.3	-1.3
2008	-1.4	-1.3	-1.1	-0.9	-0.7	-0.5	-0.4	-0.3	-0.3	-0.4	-0.6	-0.7
2009	-0.7	-0.6	-0.4	-0.1	0.2	0.4	0.5	0.5	0.6	0.9	1.1	1.3
2010	1.3	1.2	0.9	0.5	0.0	-0.4	-0.9	-1.2	-1.4	-1.5	-1.4	-1.4
2011	-1.3	-1.0	-0.7	-0.5	-0.4	-0.3	-0.3	-0.6	-0.8	-0.9	-1.0	-0.9
2012	-0.7	-0.5	-0.4	-0.4	-0.3	-0.1	0.1	0.3	0.3	0.3	0.1	-0.2
2013	-0.4	-0.4	-0.3	-0.2	-0.2	-0.2	-0.3	0.3	-0.2	-0.3	-0.3	-0.3
2014	-0.5	-0.5	-0.4	-0.2	-0.1	0.0	-0.1	0.0	0.1	0.4	0.5	0.6
2015	0.6	0.5	0.6	0.7	0.8	1.0	1.2	1.4	1.7	2.0	2.2	2.3
				LA NINA		NORMAL		ELNINO				

Source: CPC/NCEP/NOAA (2016).

The events that occurred in 2006 and 2013 were among the years with the highest annual rainfall volumes, but the ENOS phenomenon was in neutrality, indicating that there was the influence of another pattern for the marked variation in rainfall in these years. According to Souza and Nobre (1998), the TSM of the Atlantic Ocean interferes with the occurrence of rainfall in the Amazon, where the meridional and inter-hemispheric thermal gradient over the Equatorial Atlantic influences the north-south displacement of the ITCZ, which is the main meteorological system inducing rainfall in the northeast of the Amazon.

In the years 2007-2008, CPTEC/INPE (2010) identified a strong LN event, as shown by the negative IOS in Table 1. However, in 2007, the rainfall scenario was not very different from 2011 and 2012, as the rainfall in Belém, shown in Figure 9, was lower, as it should be in LN years, so it can be said that this phenomenon had little influence on the rainfall regime in the city of Belém. This region also experiences the occurrence of mesoscale atmospheric systems, such as SCMC, which are easily observed by satellite. In 2011, the year of LN (Figure 9), the larger SCMC were more frequent and restricted to the mouth of the Tocantins

River and Marajó Bay, as analysed by Sodré (2013). Although ENOS plays an important role in the interaction between TSM and precipitation in the Amazon, it is not the only modulator of precipitation in this region, which is why not all events of higher/lower precipitation are associated with this pattern.

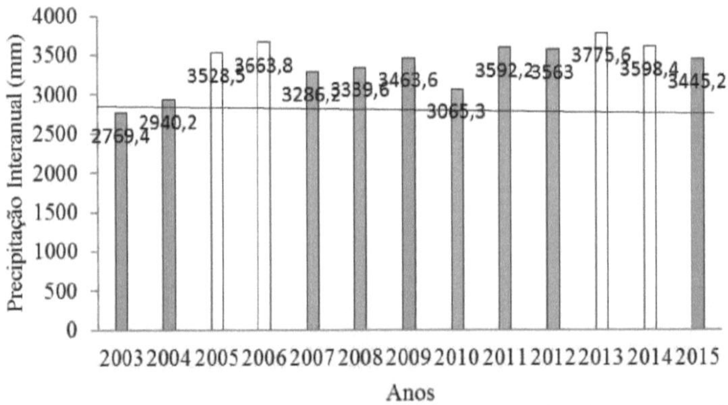

Figure 8: Interannual variability of rainfall in Belém for 2003-2015. The red bars represent the years that were under the influence of the EI Nino phenomenon, the blue bars were under the influence of the La Nina phenomenon and the white bars were years that were not under the influence of ENOS. The continuous line is the average annual rainfall.

Analysing the average monthly rainfall in mm. day -1 in the years 2003-2015 (Figure 9), it can be seen that in the months from December to May, the wettest months, there was an average monthly rainfall of 13.2 mm. day -1 (72.7% of the total).
total monthly average). The months with the highest rainfall were March and April, the monthly average for these two months was 16.2 mm. day -1 of rain (29.7 % of the total monthly average). While in the less rainy months (June to November) the monthly average was 5 mm. day -1 , which represented 27.3 % of the total monthly average.

The variability of rainfall between the two periods (rainy and less rainy) shows the influence of meso- and large-scale phenomena on the increase and decrease of rainfall. The volume of rainfall in the rainy season is related to the period when the ITCZ and MJO are active in the region (DE SOUZA and AMBRIZZI, 2006). In the less rainy period, the ILs are responsible for around 45% of the rainfall (COHEN et al., 1995).

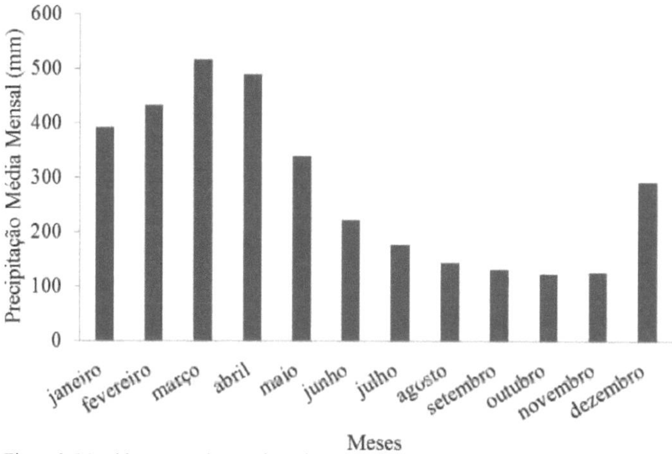

Figure 9: Monthly average in mm day-1 between 2003 and 2015 for Belém-PA.

Figure 10 shows the frequency of rainfall per hour from 2003 to 2015 in Belém for each month of the first half of the year (rainy). The figure shows that February and March had the highest number of rainy days, followed by January, April and May, and finally June. However, a major highlight of the figure is the concentration of rainfall in the afternoon until early evening for all the months considered. The dawn is the period with the lowest frequency in all the months of the first half of the year. There is also a gradual increase in the number of rain events in the early morning.

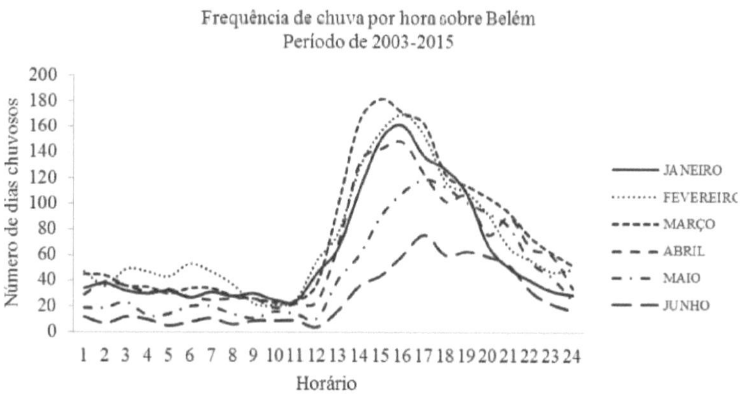

Figure 10: Hourly rainfall frequency for the months of January to June, for the period 2003 to 2015, for the Belém-PA weather station.

Figure 11, similar to the previous figure, shows the frequency of rainfall per hour for the second half of the year (less rainy, or dry). There is a reduction in the number of rainfall events compared to the first half of the year, but to a lesser extent, and most of the rainfall is also concentrated in the afternoon and evening. In the early hours of the morning, there is a noticeable reduction in rainfall, a pattern very similar to the rainy season. However, one highlight is the concentration of rainfall in the month of December, when the rains begin to return to the afternoon, and the month of July, when the greatest concentration is in the evening.

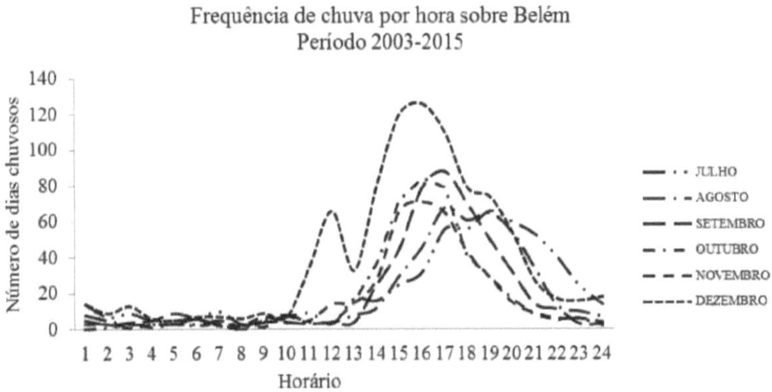

Figure 11: Hourly rainfall frequency for the months of July to December, for the period 2003 to 2015 for the Belém-PA weather station.

4.1.6 Application of Cluster Analysis

Cluster analysis was used to group data elements based on their similarity. The groups were determined in such a way as to obtain homogeneity between them and heterogeneity, thus separating the 25 stations into groups. Figure 12 shows the dendrogram representing the homogeneity between the rainfall at each station.

A horizontal cut, delimiting the formation of three intra-homogeneous groups, proved to be more effective and better distributed than one with two groups. Since two groups would not be interesting for the objective we wanted to achieve, which was to analyse the stations that had a similar rainfall distribution to Belém. So the best characterisation for Ward's method was 3 groups. A cut with a larger number of groups would have groups with only two stations.

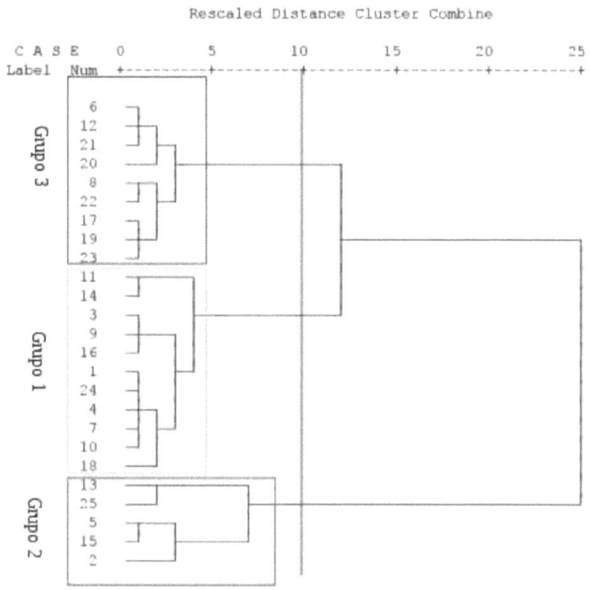

Figure 12: Dendrogram resulting from Ward's method for the 25 rain gauge stations used.

Figure 13 shows the geographical distribution of the groups in the state of Pará, where Group 1 (pink diamonds) is located in the north-west, followed by Group 2 (green cross) in the north-east and Group 3 (blue balls) in the south-east/south. It should be noted that some areas in the north-west and south of Pará, where there is relatively steep topography, difficult access conditions and/or indigenous reserves, the density of stations is relatively lower, which is why there are no stations in the white areas.

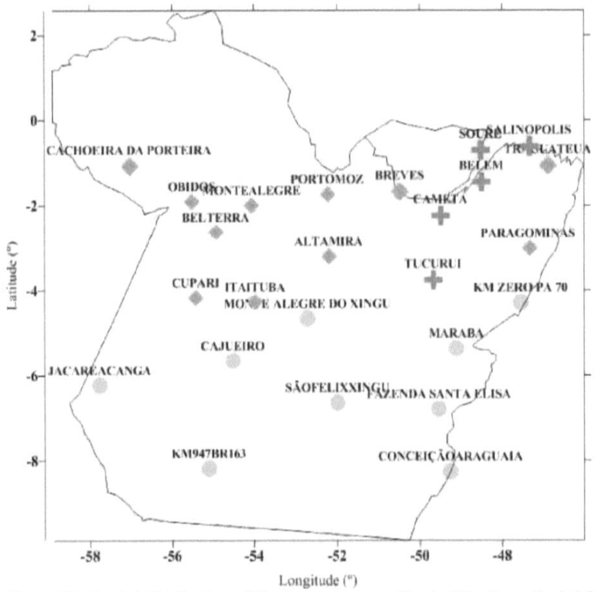

Figure 13: Spatial distribution of the 3 groups according to Ward's method. Blue ball (Group 1), green cross (Group 2) and red diamonds (Group 3).

Analysing Figure 14, we can see the similarity and difference in the monthly rainfall configuration. The similarity between the groups can be seen in the rainiest months, March and April, and the difference between them is only in the intensity of the rainfall, a profile observed in all the groups.

In Group 2, the average values are higher in the rainiest months. The same analysis can be made for the similar configuration in Group 1. The figure shows that the lowest values are between August and September, but with much lower values than in Group 2.

The relationship between Groups 1 and 3 shows that the average monthly values in Group 3 are lower than those in Group 1, especially in the months of June, July and August, which have average values of less than 100 mm in Group 3.

Thus, we can understand that the most significant meteorological systems influencing rainfall in the east of Pará (Group 2) are the ITCZ and the MFL, in addition to the influence of the rivers and their tributaries throughout the state, where the average monthly rainfall varies between 50 mm and 500 mm. In Groups 1 and 3, the system that best justifies this behaviour is the convergence of humidity from the Amazon rainforest and the North Atlantic, which is responsible for the humid climate of the northern region.

Another system responsible for the humidity in the southeast of Pará (Group 3) is the ZCOU (Humidity Convergence Zone in the southeast/northwest direction), another system that interferes with precipitation throughout the state of Pará is the SCM, which acts during the summer of HS, in addition to local systems.

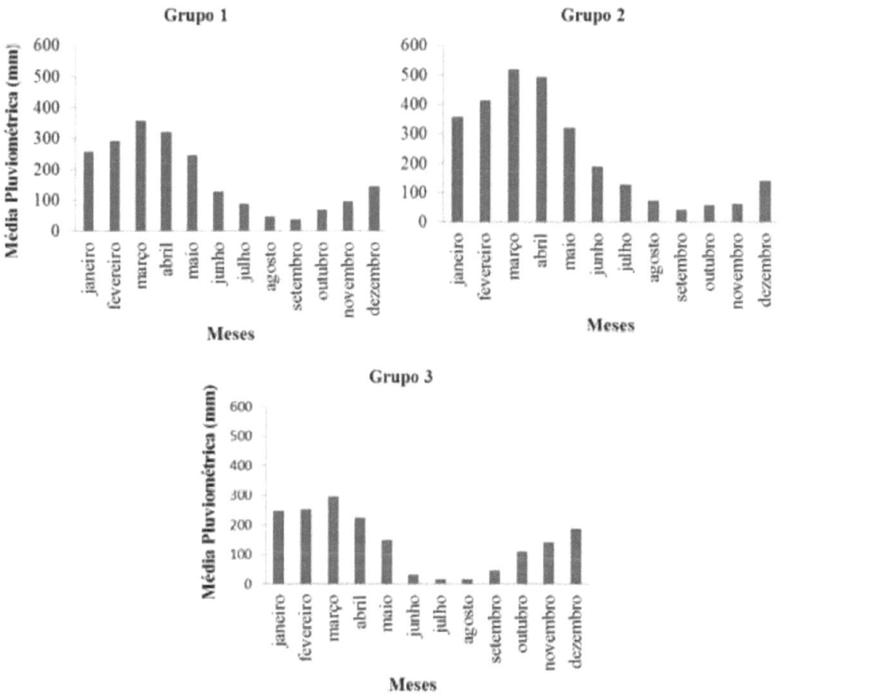

Figure 14: Histogram of the temporal distribution of rainfall by homogeneous region in the state of Pará from 2008 to 2015.

4.1.7. Homogeneous Precipitation Regions

Figure 15, representing Group 2 (Belém, Souré, Cametá, Salinópolis and Tucuruí), shows the similarity in average rainfall between the five (5) stations selected from 2008 to 2015. This similarity was only possible because the stations are located close together and have very similar rainfall patterns. In fact, the average rainfall over the course of a year is very similar between the stations. However, they differ in terms of position, scale and shape, and another differentiating characteristic is the probability of drought periods occurring.

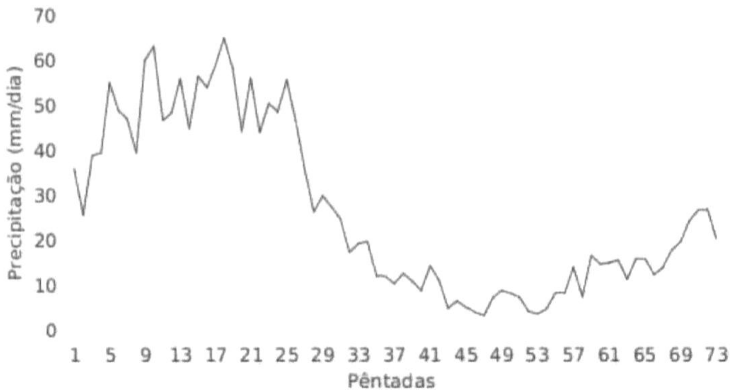

Figure 15: Average rainfall per cent according to the homogeneous area of Group 2.

Figure 16 shows the diurnal variation in rainfall at the Belém, Souré, Cametá, Salinópolis and Tucuruí stations between 2008 and 2015. When comparing the five stations with the previous study, it was observed that among the stations, only in Belém does the rainfall have a continental characteristic and is typical of the Amazon sub-regions, as the Belém station stands out in the figure with a maximum peak at 16h00 HL. This result agrees with the findings of Mota (2003), who suggested, in 3 years of TRMM (Tropical Rainfall Measuring Mission) radar rainfall data, that in the region from the Amazon River Delta to Marajó Island, the afternoon rainfall maximum is mainly due to systems of small to medium intensity and/or size, which are formed by solar heating of the surface and the surrounding air. This study also suggested that the secondary nocturnal maximum has a greater contribution from SCMs than from CCMs.

At the Souré station, a maximum is observed at night, between 03:00 and 06:00 HL, and secondary maxima in the afternoon and early evening. The existence of two maxima at Souré was investigated by Kousky (1980), who suggested that the nocturnal maximum occurs from January to May and the secondary maximum from June to September. Thus, it is suggested that the nocturnal maximum is greatly influenced by the ITCZ and the afternoon maximum is modulated by the action of the breeze and the formation of LI in the region. These differences in maximum rainfall values suggest that the breeze, LI and SCM systems associated with large-scale mechanisms, such as the ITCZ, contribute to the spatial and temporal variability of rainfall in Souré.

The Cametá station showed only one maximum in the diurnal cycle, occurring at night (18h00 to OOhOO HL) and a minimum occurring in the early hours of the morning. The figure shows that Cametá's rainfall increases at night, reaching values of between 0.2 and 0.5 mm/h during this period. It can be seen that the rainfall in Belém and Cametá is similar in the early morning. A pioneering study by Asnani (1993) proposes that rainfall in the Amazon has two conceptual models: the continental model, with rainfall occurring preferentially in the afternoon and early evening, as shown in the figure, where Belém has more rainfall in the afternoon and early evening; and the maritime model, with rainfall occurring predominantly at night, as observed in Cametá. Rainfall in Cametá is also heavily influenced by the BF, as the city is surrounded by rivers and forests and also suffers the influence of the LI, intensified by the convergence associated with the ITCZ, which favours the occurrence of rain, especially in the late afternoon and early evening.

The diurnal cycle in Salinópolis had the least amplitude of the four stations considered. There is a maximum in the early hours of the day and secondary peaks in the afternoon and evening. The north-eastern region of Pará, more specifically Salinópolis, is influenced by various global-scale systems, including the EI Nino and La Nina, and local-scale systems, such as the breeze effect, which cause rainfall rates to decrease and increase (MÁCOLA and EL-ROBRINI, 2004). Although Salinópolis is in a coastal region, rainfall throughout the day is low, which can also be explained by the intensification of urban interventions as a way of commercialising the city. The diurnal cycle of Souré and Salinópolis appears to be maritime. According to Oliveira (1986), the atmospheric system that directly influences local precipitation is the ITCZ, mainly in the coastal region of the Amazon, reaching around 500 kilometres into the continent, and Souré is one of the regions affected by this system, as it is located close to the coast. Therefore, it is speculated that the coastal locations have a nocturnal maximum associated with maritime circulations and the influence of the ITCZ, while the Belém station has a maximum in the afternoon, possibly resulting from solar heating of the surface and adjacent air.

The Tucuruí station, on the other hand, showed two maxima in the diurnal cycle. The main one at dawn and a secondary one between 19h00 and 23h00 HL. The main minimum of the diurnal cycle occurred between 07h00 and 18h00 HL. Tucuruí is typically continental, and the main maximum can be explained by the maximum convection over the region, which generally occurs in the late afternoon and early evening, in response to daytime surface heating (SUI et al., 1997). On the other hand, the nocturnal maximum is associated with the LI entering the continent. This can be seen in the figure in the displacement of the breeze when it begins in Belém, where precipitation occurs in the early afternoon, followed by Cametá, with precipitation occurring from 8 to 11 p.m., and finally Tucuruí, with precipitation occurring from 11 p.m. to 2 a.m. The maximum rainfall during the night is associated with the LI entering the continent. The maximum rainfall at night in this region can also be explained by the formation of SCMs (MOTA, 2003), as well as contributions associated with the onset of synoptic systems at the end of spring and summer, such as the SACZ (KODA-MA, 1992; CARVALHO et al., 2004; FERREIRA et al., 2004). However, Silva (1999) observed that MCCs are also responsible for this rainfall cycle in the region.

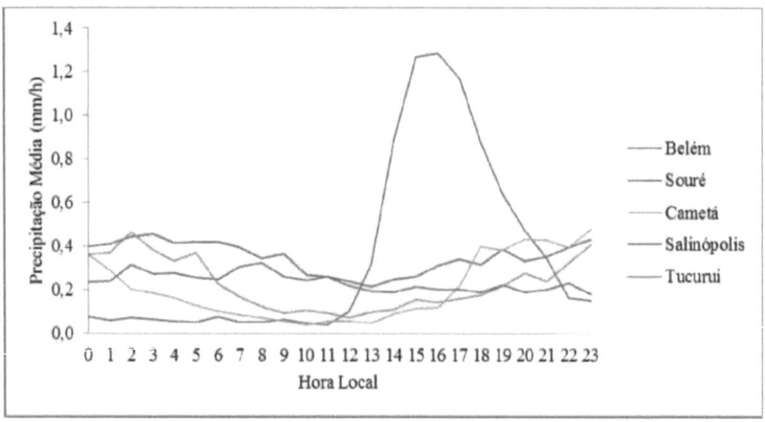

Figure 16: Representation of the diurnal variation in precipitation for Group 2 from 2008 to 2015.

CHAPTER 5

CONCLUSIONS

From the results obtained in this study, it can be concluded that:

The "famous afternoon rain" in Belém makes a major contribution to the monthly totals, with 4pm HL being the time of greatest contribution. In the period between dawn and early morning, the contribution values are minimal.

The contribution in the rainiest months (December to May) was 26 per cent of the monthly total at 16:00 HL and the lowest was in May, with 14 per cent. It can be seen that from December to March this contribution is restricted to the afternoon. In the months of April and May, it can be seen that they increase a little in the evening. This is still seen in the less rainy months (June to August) and returns to the afternoon from September to November. This change is related to the action of the ITCZ and mesoscale effects, such as LI associated with BM and also the formation of BF.

The increase in contributions at night is due to the northward shift of the ITCZ and the formation of SCMs, as well as MLFs with greater frequency in August, as well as local effects such as BM and DOL.

The years 2006 to 2013 were among the years with the highest and lowest annual rainfall, respectively, but in these years the ENOS phenomenon was in neutral, indicating the influence of another pattern, such as the Atlantic Ocean's SST.

Rainfall events in Belém increase in frequency between the afternoon and early evening, in any month of the year. The system that most influences this pattern is the LI associated with the BM.

Group 1 is located in the north-west and its rainfall is mainly influenced by the north-south displacement of the ITCZ, Group 2 is located in the north-east of the state with rainfall from local convection and the LI and MLF associated with the BM and finally Group 3 occupies the south-central part of the state with rainfall subject to the influence of frontal systems, the ITCZ and local convection.

Finally, after analysing Group 2 to identify regions in the state of Pará with rainfall similar to Belém in the afternoon, it can be said that Belém has a rainfall regime that differs from the other municipalities, being the only municipality to show high rainfall in the afternoon in all months, compared to the others.

CHAPTER 6

BIBLIOGRAPHICAL REFERENCES

ACEITUNO, P. On the Functioning of the South American Sector. Part 1: Surface Climate. **Monthly Weather Review.** Volume 116, 1988.

NATIONAL WATER AGENCY (ANA). Hydrological information system (HidroWeb). Available at : < http://hidroweb.ana.gov.br/ > . Accessed on 10 September 2016.

ALBUQUERQUE, M.F.; SOUZA, E.B.; OLIVEIRA, M.C.F.; SOUZA JR, J.A. Precipitation in the mesoregions of the State of Pará: Climatology, variability and trend in recent decades (1978-2008). Brazilian Journal of Climatology. ISSN: 1980-055X, v.6, 2010.

ALCÂNTARA, C. R.; SILVA DIAS, M. A. F.; SOUZA, E. P.; COHEN, J. C. P. Verification of the Role of the Low Levei Jets in Amazon Squall Lines. **Atmospheric Research,** v. 100, p. 36-44, 2011.

ALCÂNTARA, C. R.; SOUZA, E. P.; SILVA DIAS, M. A. F.; BIAZETO, B. Influence of medium and low level jets on cloud processes: Numerical study of an Amazonian instability line. **Revista Brasileira de Meteorologia,** v. 29, n. 1, 29-46, 2014.

AMANAJÁS, J. C.; BRAGA, C. C. 2012. Spatio-temporal rainfall patterns in the Eastern Amazon using Multivariate Analysis. **Revista Brasileira de Meteorologia,** v.27, n.4, 323 - 338.

ASNANI, G.C. Tropical Meteorology. Pune, India: Noble Printers, 2v. 121P, 1993.

ATKINSON, B.W. Meso-scale atmospheric circulation. London Academic Press, 495p., charp. 5 e 6. 1981.

BARBOSA, E. R.; Associations between the subtropical jet stream and mesoscale convective complexes. Part I: The influence of the subtropical jet stream on the development of mesoscale convective complexes. In: **XI Brazilian Congress of Meteorology,** Rio de Janeiro. SBMET, 2000.

BARBOSA, R. L. **Interaction of convective disturbances initiated on the north coast of Brazil with easterly wave disturbances.** 81 p. Master's dissertation from the Postgraduate Programme in Meteorology - National Institute for Space Research, São José dos Campos, 2005.

BARBOSA, R. L.; OYAMA, M. D.; MACHADO, L. A. Climatology of convective disturbances initiated on the northern coast of Brazil. **Revista Brasileira de Meteorologia,** V. 21, n. 1, p. 107-117, 2006.

BARROS, V. R.; GRIMM, A. M.; DOYLE, M. E. Relationship between temperature and circulation in Southeastern South America and its influence from EI Nino and La Nina events. **Journal of the Meteorological Society of Japan,** v. 80, p. 33-44, 2002.

BARROS, S.S. Precipitation at the Alcântara Launch Centre: observational and modelling aspects. Master's dissertation in Meteorology. National Institute for Space Research, São José dos Campos, 112p. (INPE-15319- TDI/1362), 2008.

BASTOS, T. X.; PACHECO, N.A.; NECHET, D.; ABREU, T. D. Aspectos Climáticos de Belém nos Últimos 100 anos, Boletim da Pesquisa e Desenvolvimento. **Documents.** EMBRAPA-CPATU, Belém-Pará, 2002.

BAUMBACH, G.; VOGT, U. Experimental determination of the effect of mountain- valley breeze circulation on air pollution in the vicinity of Freiburg. **Atmospheric Environment,** 33, 4019-4027, 1999.

BERBERY, E. H.; BARROS, V. R. The Hydrologic Cycle of the La Plata Basin in South America. **Journal of Hydrometeorology,** v. 3, p. 630-645, 2002.

BERRI, G. J. et al. The influence of ENOS in the flows of the Upper Paraná River of South America over the past 100 years. **Journal of Hydrometeorology,** v.3, p.57-65, 2002.

BETTS, G.M.; GROWERM. R.W.; MONCRIEFF, W. Structure and Motion of the Tropical Squall Lines over Venezuela. **Quart. J. Roy Meteor.** Soc., 102, 395-404, 1976.

CAETANO, J. **Analysis of easterly waves over the east coast of north-eastern Brazil for the period 1999-2009.** Master's dissertation. Postgraduate Programme in Meteorology at the Geoscience Institute of the Mathematics and Nature Centre of the Federal University of Rio de Janeiro (PPGM-IGEO-CCMNUFRJ), 2011.

CAMILLONI, I. A.; BARROS, V. R. The Paraná River response to EI Nino 1982-83 and 1997-98 events. **Journal of Hydrometeorology,** v.l, p.412-430, 2000.

CARVALHO, L. M. V; JONES, C.; LIEBMANN, B. Extreme precipitation events in Southeastern America and large-scale convective patterns in the South Atlantic Convergence Zone. **Journal of Climate,** v.15, p.2377-2394, 2002.

CARVALHO, L.M.V; JONES, C.; LIEBMANN, B. The South Atlantic Convergence Zone: persistence, intensity, from, extreme precipitation and relationships with intraseasonal activity. **Journal Climate,** 17, p. 88-108, 2004.

CAVALCANTI, I. F. A. A **study on interactions between synoptic-scale circulation systems and local circulations.** MSc thesis. INPE-2494-TDL/097, São José dos Campos, Brazil, 1982.

CAVALCANTE, I.F.A.; FERREIRA, N.J.; SILVA DIAS, M.A.F.; JUSTI, M.G.A. Weather and climate in Brazil. São Paulo: **Oficina de Textos,** 2009.

COELHO, C. A. et al. The 2014 Southeast Brazil austral summer drought: regional scale mechanisms and teleconnections. Climate Dynamics, **Springer,** p. 1-16, 2015.

COELHO, M.S.; GAN, M.A.; CONFORTE, J.C. Study of the variability of the position and cloudiness associated with the Atlantic ITCZ during the rainy season of 1998 and 1999 in Northeast Brazil. **Revista Brasileira de Meteorologia,** v.19, n.l, p.23- 34, 2004.

COHEN, J.C.P.; SILVA DIAS, M.A.F.; NOBRE, C.A. Aspectos climatológicos das linhas de instabilidade na Amazônia. **Climanálise - Bulletin of Climate Monitoring and Analysis,** 4_(11): 34 - 40, 1989.

COHEN, J. C. P.; SILVA DIAS, M. A. F.; NOBRE, C. Environmental conditions associated with Amazonian squall lines: A case study. **Mon. Wea. Rev.,** 123, 3163- 3174,1995.

COHEN, J. C. P. **Propagation Mechanism and Structure of Instability Lines in the Amazon.** Doctoral Thesis in Meteorology. São Paulo, IAG- USP, 1996.

COHEN, J.C.P, FITZJARRALD, D.R.; D'OLIVEIRA, F.A.F.; SARAIVA, L; BARBOSA, I.R.S.; GANDU, A.W.; KUHN, P.A. Radar-observed spatial and temporal rainfall variability near the Tapajós-Amazon confluence. **Revista Brasileira de Meteorlogia;** 29 (n.esp.): 23-30, 2014.

COLLISCHONN, W.; TUCCI, C. E.; CLARKE, R.T. Further evidence of changes in the hydrological regime of the River Paraguay: part of a wider phenomenon of climate change? **Journal of Hydrology,** v. 245, p. 218-238, 2001.

CORNISH, R. **Cluster Analysis.** Mathematics Learning. Support Centre. 2007.

COUTINHO, E. C.; FISCH, G. Eastern Wave Disturbances (EWDs) in the region of the Alcântara Launch Centre - MA. **Revista Brasileira de Meteorologia,** v.22, n.2, p. 193-203, 2007.

CHOU, S.C. **Analysis of Eastern Wave Disturbances over the South Equatorial Atlantic Ocean.** 153 p. (INPE - 5222 - TDL / 437). Dissertation (Master's Degree in Meteorology) - National Institute for Space Research, São José dos Campos, 1990. Available at: . Accessed on: 15/01/2017.

CLIMANÁLISE, **Climate Analysis Monitoring Bulletin.** São Paulo: SCT/INPE-CPTEC, v. 25, n. 01, January 2010.

CPC/NOAA. **Cold & Warm Episodes by Season.** Climate Prediction Center Internet Team, 04 May 2010. Available at: Accessed 06 Jun 2016.

DE SOUZA, E.B.; AMBRIZZI, T. Modulation of the intraseasonal rainfall over tropical Brazil by the Madden-Julian oscillation. **International Journal of Climatology.** São Paulo, v.26, p. 1759-1776, 2006.

DE SOUZA, E. B.; KAYANO, M. T.; AMBRIZZI, T. The regional precipitation over the eastern Amazon/Northeast Brazil modulated by tropical Pacific and Atlantic SST anomalies on a weekly timescale. **Revista Brasileira de Meteorologia** (Impresso), v. 19, n.3, p. 113-122, 2004.

DE SOUZA, E.B.; NOBRE, P. A review of the Dipole Pattern in the tropical Atlantic Ocean. **Revista Brasileira de Meteorologia,** v. 13, n. 1, p. 31-44,1998.

DUNN, G.E. Cyclogenesis in the Tropical Atlantic. **Bulletin American Meteorological Society,** 21, p.215-229, 1940.

DUQUIA, C. G. E SILVA DIAS, M. A. F. Mesoscale convective complex: a case study for western Rio Grande do Sul. In: VIII CBMet, SBMET. **Proceedings,** 2: 610-612, 1994.

DRUMOND, A. R. M.; AMBRIZZI, T. The role of SST on the South American atmospheric circulation observed during the 2000/2001 Austral Summer. **Climate Dynamics,** v. 24, 781-791, 2005.

DRUMOND, A.; MARENGO, J.; AMBRIZZI, T.; NIETO, R.; MOREIRA, L.; GIMENO, L. The role of the Amazon Basin moisture in the atmospheric branch of the hydrological cycle: a Lagrangian analysis. **Hydrology and Earth System Sciences,** v. 18, p. 2577-2598, 2014.

ESPINOZA, E.S. **Disturbances in the easterly winds in the Tropical Atlantic.** Master's dissertation in Meteorology-National Institute for Space Research, São José dos Campos. 149p. (INPE-6347-TDL/598), 1996.

EVERITT, B.S.; DUNN, G. **Applied Multivariate Data Analysis.** Edward Arnold, London, 1991.

EVERITT, B.S. **Cluster analysis.** Heinemann Educational Books, London: Academic Press, 3ª edition, p.170, 1993.

FERREIRA, N. J.; SANCHES, M.; SILVA DIAS, M. A. F. Composition of the South Atlantic Convergence Zone in periods of El Nino and La Nina. **Revista Brasileira de Meteorologia,** 19(1): 89-98. 2004.

FERREIRA, D. B da S.; SOUZA, E. B.; MORAES, B.C. 2013. Hourly cycle of precipitation in eastern Amazonia during the rainy season. **Brazilian Journal of Climatology.** V. 13, p. 74-86, 2013.

FERREIRA, D. B. S.; DE SOUZA, E. B.; DE MORAES, B. C.; FILHO, L. G. M. Spatial and Temporal Variability of Rainfall in Eastern Amazon during the Rainy Season. **The Scientific Word Journal,** Volume 2015. Article ID 209783, 9 pages, 2015.

FIGUEIREDO, J.C. and SCOLAR, J. Study of the trajectory of mesoscale convective systems in South America. **VII Congreso Argentino de Meteorologia and VII Congreso Latinoamericano e Ibérico de Meteorologia**, Buenos Aires, Argentina, September 1996,165-166. 1996.

FIGUEROA, A. S. N.; NOBRE, C. Precipitation distribution over central and estern Tropical South America. **Climanálise,** v. 5, n. 6, p. 36-45, 1990.

FIGUEIREDO, C.T.; MOTA, M.A.S. Analysis of precipitation in the metropolitan region of Belém (PA) over the course of a year (August/2008 to July/2009). XVI Congress of Meteorology. **Proceedings.** 2010.

FIGUEIREDO, J.C. and SCOLAR, J. Study of the trajectory of mesoscale convective systems in South America. **VII Congreso Argentino de Meteorologia and VII Congreso Latinoamericano e Ibérico de Meteorologia**, Buenos Aires, Argentina, September 1996, 165-166. 1996.

FILHO, T. F.; ASSAD, E. C.; LIMA, P. R. S. R. Homogeneous rainfall regions in Brazil. **R. Pesq. Agropec. Bras.,** 40: 311-322, 2005.

FILHO, A. J. P. et al. Convective rainfall in Amazonia and adjacent tropics. Atmospheric and

Climate Sciences, **Scientific Research Publishing,** v. 5, n. 02, p. 137, 2015.

FISCH, G.; MARENGO, J.A.; NOBRE, C.A. Clima da Amazônia. **Climanálise 10 anos.** CPTEC/INPE. 1997.

FISCH, G.; MARENGO, J. A.; NOBRE, C. A. A general review of the climate of Amazonia. **Acta Amazônica,** V. 28, p. 101-106, 1998.

FRAGOSO, M.; GOMES, P. T. Classification of daily abundant rainfall patterns and associated large-scale atmospheric circulation types in Southern Portugal. International **Journal of Climatology,** 28:537-544, 2008.

FREITAS, S. R.; LONGO, K. M.; SILVA DIAS, M. A. F.; CHATFIELD, R.; SILVA DIAS, P. L., ARTAXO, P.; ANDREAE, M.; GRELL, G.; RODRIGUES, L. F.; FAZENDA, A.; PANETTA, J. The Coupled Aerosol and Tracer Transport model to the Brazilian developments on the Regional Atmospheric Modeling System (CATT- BRAMS). Part 1: Model description and evaluation. **Atmospheric Chemistry and Physics Discussion,** v.7, p. 8571-8595, 2007.

FREITAS, A. C.V.; AÍMOLA, L.; AMBRIZZI, T.; OLIVEIRA, C. P. Extreme Intertropical Convergence Zone shifts over Southern Maritime Continent. **Atmospheric Science Letters.** 18: 2-10, 2017. DOI: 10.1002/asl.716.

FRIEDMAN, H. P.; RUBIN, J. On some invariant criteria for grouping data. **Journal of the American Statistical Association,** 62,1159-1178,1967.

FITZJARRALD, D. R; SAKAI, R. K, MORAES OL, COSME DE OLIVEIRA R, ACEVEDO OC, CZIKOWSKY MJ, BELDINI T. Spatial and temporal rainfall variability near the Amazon-Tapajós confluence. **Journal of Geophysical Research.** Biogeosciences 113(G1): G00BU, 2008 doi: 10.1029/2007JG000596.

GANDU, A. W.; SILVA DIAS, P. L. Impact of tropical heat sources on the South American tropospheric upper circulation and subsidence. **Journal of Geophysics Research,** v. 103, p. 6001-60015, 1998.

GAMACHE, J. F.; HOUZE JR, R. A. Mesoscale air motions associated with a tropical squall line. **Monthly Weather Review,** V. 110, p. 118-135, 1982.

GARSTANG, M., GRECO, S.; SCALA, J.; SWAP, R. ULANSKI, S. FITZJARRALD, D.; MARTIN, D.; BROWELL, E.; SHIPMAN, M.; CONNORS, V; HARRISS, R. TALBOT, R. The Amazon Boundary Layer Experiment (ABLE 2B): A meteorological perspective. **Bulletin of the American Meteorological Society,** v. 71, p. 19-32, 1990.

GONG, X.; RICHMAN, M. R. On the application of cluster analysis to growing season precipitation data in North American East of the Rockies. **Journal of Climate,** 8:897-924, 1995.

GORE JR, P. A. Cluster Analysis. Handbook of Applied Multivariate Statistics and Mathematical Modelling. **Department of Psychology, Southern Illinois University**, Carbondale, Illinois, ch. 2, p. 297-321. 2000.

GREC/IAG-USP. **Monthly climate report. In: Climate Studies Group, GrECs/IAG-USP.** University of São Paulo, USP, 2014.

GREEN, M. C.; FLOCCHINI, R. G.; MYRUP, L. O. Use of temporal principal component analysis to determine seasonal periods. **Journal of Applied Meteorology,** 32:986-995, 1993.

GRIMM, A. M.; BARROS, V. R.; DOYLEe, M. E. Climate variability in Southern South America associated with EI Nino and La Nina events. **Journal of Climate,** 13: 35-58, 2000.

GRIMM, A.M. The EI Nino impact on the summer monsoon in Brazil: Regional processes versus remote influences. **Journal of Climate,** v. 16, p. 263-280, 2003.

GRIMM, A. M. How do La Nina events disturb the summer monsoon system in Brazil? **Climate Dynamics,** 22: 123-138, 2004.

GRIMM. A. M.; TEDESCHI, R. G. ENSO and extreme rainfall events in south american. **Journal of Climate.** V.22. American Meteorological Society. 2009. DOI: 101175/2008JCLI2429.1.

GRIMM, A. M. Interannual climate variability in South America: impacts on seasonal precipitation, extreme events, and possible effects of climate change. **Stochastic Environmental Research and Risk Assessment,** 25(4): p.537-554, 2011.

GUEDES, R. L., MACHADO, L. A. T, SILVEIRA, J. M. B., ALVES, M. A. S. E WALTZ, R. C. Trajectories of convective systems over the American continent. In: VIII CBMet. **Proceedings,** 2: 77-80,1994.

GUEDES, R.L. **Mesoscale Systems.** In: Course in Image Interpretation and Meteorological Analysis. University Extension Course. São José dos Campos: Vale do Paraíba University, 28p. 1985.

GUEDES, R. V. S.; LIMA, F. J. L.; AMANAJÁS, J. C. BRAGA, C. C. Principal component analysis of rainfall in the state of Piauí and clustering by Ward's method. **Journal of Geography.** Recife, UFPE- DCG/NAPA, V. 27, n. 1, 2010.

HAIR, J. F; BACK, R. E; TATHAM, A. R. L. **Multivariate Data Analysis.** Bookman, 5. ed., ISBN 8536304820, 9788536304828. Porto Alegre, 2005.

HAMILTON, R. A.; ARCHBOLD, J. N. Meteorology of Nigeria and adjacent territory. **Quart. J. Roy. Meteor.** Soc., 71, 231-262, 1945.

HASTENRATH, S.; HELLER, L. Dynamics of climatic hazards in Northeast Brazil. **Quart. J. Roy. Meteor Soc.,** V 103, n. 435, p. 77-92,1977.

HASTENRATH, S. The intertropical convergence zone of the eastern pacific revisited. **International Journal of climatology.** v. 22, p. 347-356, 2002. Published Online in Wiley InterScience Q. DOI: 10.1002/joc.739.

HORN, D. **A study of personality syndromes.** Character and Personality, 12, 257- 274,1943.

HOUZE, R. A.,JR. Structure and dynamics of a tropical squall-line system observed during

GATE. **Mon. Wea. Rev.,** 105, 1540-1567,1977.

HOUZE, R. A. Mesoscale convective systems. **Reviews of Geophysics,** Wiley Online Library, v. 42, n. 4, 2004.

IBGE. Demographic Census, 2010.

INMET, Climatological Normals (1961-1990), INMET, Brasília-DF, 1992.

INNOCENTINI, V.; ARANTES, F. O.; FERREIRA, R. J. and MICHELETO, R. G. Maritime Agitation on the Northeast Coast of Brazil Associated with East African Disturbances. **Revista Brasileira de Meteorologia.** V. 20, n. 3, p. 367-374, 2005.

JOHNSON, S. C. **Hierarchical clustering schemes.** Psychometrical, 32, 241-254, 1967.

KAYANO, M. T. **A climatological and synoptic study using radiosonde data (1968-1976) from Manaus and Belém.** 111 f. (INPE-1559-TDL/013). Dissertation (Master's in Meteorology) - INPE, São José dos Campos, 1979.

KAYANO, M T; RAO, V B; ANDREOLI, R V. "A review of short - term climate variability mechanisms", **Advances in Space Research** , v. 35, p. 843 - 851, 2005.

KAYANO, M.T; ANDREOLI, R.V. Relationships between rainfall anomalies over northeastern Brazil and the EI Nin o Southern Oscillation. **Journal of Geophysical Research,** 111: D13101, 2006.

KOUSKY, V. E. Frontal Influences on Northeast Brazil. **Monthly Weather Review,**v. 107, p. 1140-1153, 1979.

KOUSKY, V.E. Diurnal rainfall variation in the northeast Brazil. **Mon. Wea. Rev.** 108, 488-498, 1980.

KOUSKY, V.E.; KAGANO, M.T. A climatological study of the tropospheric circulation over the Amazon region. **Acta Amazon.** v.ll, n.4, p.743-753, 1981.

KOUSKY, V.E.; MOLION, L.C.B. **Uma contribuição à climatologia da dinâmica da troposfera sobre a Amazônia.** São José dos Campos. Space Research Institute, 1981.

KOUSKY, V. E.; CAVALCANTIavalcanti, I. F. A., and Kayano, M. T.: A review of the Southern Oscillation: oceanic-atmospheric circulation changes and related rainfall anomalies, **Tellus A,** 36, 490-504,1984.

LANDIM, P. M. B. **Statistical analysis of multivariate geological data.** Oficina de Textos Publisher, 208p. 2011.

LANCE, G. N.; WILLIAMS, W. T. A general theory of classificatory sorting strategies: 1 hierarchical system. **Computer Journal,** 9, 373-380, 1967.

LAU, K.M.; CHAN, P.H.: Intraseasonal and Interannual variability of tropical convection: A possible link between the 40-50 day oscillation and ENSO? **Journal Atmospheric Sciences,** 45, p.506-521, 1988.

LAURENT, H.; MACHADO, L. A. T; HERDIES, D.; MORALES, C. Characteristics of convective systems observed by satellite during the WETAMC/LBA experiment. In: **XI Brazilian Congress of Meteorology,** Rio de Janeiro. [SI: s.n.], p. 16-20. 2000.

LIEBMANN, B.; HEBDON, H.H. Synoptic-scale disturbance near the equator. **Journal Atmospheric Sciences.** 47, 1463-1479. 1990.

LIEBMANN, B.; JONES, C.; CARVALHO, L. M. V. Interannual variability of extreme precipitation events in the State of São Paulo, Brazil. **Journal of Climate.** v.14, p.208-218, 2001.

LINDEN, R. Clustering techniques. **FSMA Information Systems Journal,** n. 4, pp. 18-36, 2009. Accessed on 21/01/2017.

LOPES, M.N.G. **Regional aspects of precipitation variability in the State of Pará: observational study and high-resolution climate modelling.** Master's dissertation in Environmental Sciences - Federal University of Pará (UFPA), Belém, 117f., 2009.

LOUREIRO, R. S.; SARAIVA, J. M. ; SARAIVA, I; SENNA, R. C. ; FREDÓ, A. S. Study of extreme precipitation events in 2009 in the state of Pará. **Revista Brasileira de Meteorologia** (Impresso), v. 29, p. 83-94, 2014.

LU, L.; DENNING, A.S.; SILVA DIAS, M.A.F.; SILVA DIAS, P.; LONGO, M.; FREITAS, S.R.; SAATCHI, S. Mesoscale circulations and atmospheric CO2

variations in the Tapajós region, Pará, Brazil. **Journal of Geophysical Research.** 110 9D21102):l-17, 2005.

MACOLA, G; EL-ROBRINI, M. Guaras Island (Mariteua) - Curuçá Municipality (NE Pará): Physical, Meteorological and Oceanographic Aspects. Final Report. Belém, **DNPM/CPRM,** 35p., 2004.

MACHADO, L. A. T.; GUEDES, R. L.; ALVES, M. A.S. Structural characteristics of convective systems and convective forcing in South America observed by satellites. **Climanálise 10 anos, CPTEC/INPE,** p. 110-122, 1997.

MACHADO, L. A. T.; LAURENT, H. The convective system area expansion over Amazonia and its relationships with convective system life duration and high-level Wind divergence. **Monthly Weather Review,** V. 132, p. 714-725, 2004.

MACHADO, L.A.T.; LAURENT, H.; DESSAY, N.; MIRANDA, I. Seasonal and diurnal variability of convection over the Amazonia: A comparison of different vegetation types and large scale forcing. **Theoretical and Applied Climatology,** v.78, n.l, p.61-77, 2004.

MACHADO, L. A. T., SILVA DIAS, M. A. F, MORALES, C., FISCH, G., VILA, D., ALBRECHT, GOODMAN, R. J, CALHEIROS, A. J. P., BISCARO, T., KUMMEROW, C., COHEN, J., FITZJARRALD, D, NASCIMENTO, E. L., SAKAMOTO, M. S., CUNNINGHAM, C, CHABOUREAU, J.P, PETERSEN, W. A., ADAMS, D. K., BALDINI, L. , ANGELIS, C. E, SAPUCCI, L. E, SALIO, P., BARBOSA, H. M. J., LANDULFO, E., SOUZA, R. A. E, BLAKESLEE, R. J.,BAILEY, J, FREITAS, S., LIMA, W. F. A.;TOKAY,A. The Chuva Project How Does Convection Vary Across Brazil? **American**

Meteorological Society .Sept. 2014.

MACHADO, L.; ROSSOW, W. Structural characteristics and radiative properties of tropical cloud clusters. Monthly Weather Review, v.121, n.12, p.3234-3260, 1993.

MADDEN, R. A.; JULIAN, P. R. Observations of the 40-50 day tropical oscillation. **Monthly Weather Review,** V. 122, n. 5, p. 814-837, May. 1994.

MADDOX, R.A. Mesoscale convective complexes. **B. Am. Meteorol. Soc.,** v. 61, p. 1374-1387, 1980.

MADDOX, Robert A. Large-scale meteorological conditions associated with midlatitude, mesoscale convective complexes. **Mon. Weather Rev.,** v. 111, n. 7, p. 1475-1493, 1983.

MARENGO, J. A. Global and regional climate change: Evaluation of Brazil's current climate and projections of future climate scenarios. **Brazilian Journal of Meteorology.** São Paulo, v. 16, p. 1-18, 2002.

MARENGO, J. A.; HASTENRATH, S. Case studies of extreme climatic events in the Amazon basin. **Journal of Climate,** v. 6, p. 617-627,1993.

MATOS, A. P; COHEN, J. C. P. Breeze circulation and the precipitation band on the eastern shore of Marajó Bay. **Ciência e Natura** 38: 21-27. 2015.

MATOS, A. P.; COHEN, J. C. P. Breeze circulation and precipitation band on the eastern shore of Marajó Bay. **Ciência e Natura,** V. 38, Ed. Especial-IX Workshop Brasileiro de Micrometeorlogia, p. 21-27, 2016.

MENDONÇA. E; DANNI-OLIVEIRAI. M. Climatologia noções básicas e climas do Brasil. São Paulo: **Oficina de Textos,** 2007.

MELO NETO, M. S. de. Soil erosion on the slopes of the Jaguaribe river basin - Conceição do Almeida - Bahia. In: CONGRESS OF RESEARCHERS OF THE SOUTHERN RECÔNCAVO OF BAHIA, 1. **Proceedings.** Amargosa, UFRB, p. 134-135, 2007a.

MELO NETO, M. S. de. Environmental degradation on the northern slopes of the Jaguaribe river basin - Conceição do Almeida - Bahia In: SIMPÓSIO BRASILEIRO DE GEOGRAFIA FÍSICA APLICADA, 12. **Proceedings.** Natal, UFRN, 2007. CD-ROM, 2007(b).

MOLION, L. C. B. Climatologia dinâmica da região Amazônica: mecanismos de precipitação. **Revista Brasileira de Meteorologia,** V. 2, p. 107-117, 1987.

MOLION, L. C. B.; BERNADO, S. O. A Review of Dynamics in the Brazilian Northeast. **Revista Brasileira de Meteorologia,** v. 17, n.l, p. 1-10, 2002.

MORAES, B. C; COSTA, J. M. N; COSTA, A. C. L; COSTA, M. H. Spatial and temporal variation of precipitation in the state of Pará. **Acta Amazônia,** v. 35, p. 207- 214, 2005.

MORAES, B. C; SÁ, L. D.A. Wavelet analysis of the precipitation series of the Caxiuanã National Forest - PA. In: **XIV Brazilian Congress of Meteorology,** Florianópolis, 2006.

MOTA, G.V. **Observational study of easterly wave disturbances in the Brazilian**

Northeast. São Paulo. Master's dissertation, 1997.

MOTA, G. V. **Characteristics of Rainfall and Precipitation Features Defined by the Tropical Rainfall Measuring Mission over South America.** Doctoral Thesis. Department of Meterorology, University of Utah, Utah, USA. 215p. 2003.

MOURA, A. D.; SHUKLA, J. On the dynamics of drougts in Northeast Brazil: Observation, Theory, and numerical experiments with a general circulation model. **J. Atmos. Sei,** V. 38, n. 12, p. 2653-2675, 1981.

MÚNNICH, M.; NEELIN, J. D. Seasonal influence of ENSO on the Atlantic ITCZ and equatorial South America. **Geophysical Research Letters,** v.20, 2005.

MUNOZ-DÍAZ, D.; RODRIGO, F. S. Spatio-temporal patterns of seasonal rainfall in Spain (1912-2000) using cluster and principal component analysis: comparison. **Annales Geophysical,** 22:1435-1448, 2004.

McQUITTY, L. L. **Pattern Analytic Clustering Lanham: University Press of America,** 1987.

NAGHETTINI, M.; PINTO, E.J.A. **Hidrologia Estatística.** Belo Horizonte: CPRM/SGB, 2007.

NECHET, D. Diurnal variability of precipitation in Belém-Pa. **Proceedings of** the 3° Brazilian Congress of Meteorology, 204-211. 1984a.

NECHET, D. Diurnal variability of precipitation in Belém-PA, In: CONGRESSO BRASILEIRO DE METEOROLOGIA, 3., 1984, Belo Horizonte. **Proceedings.** Rio de Janeiro: SBM. p.204-211, 1984b.

NECHET. D. Diurnal variability of precipitation in Belém-PA: application to medium and long-term planning. **Climatological Bulletin. Presidente Prudente,** SP, v.2, n.3, p.223-227, jul. 1997.

NETO, D. J. de S.; VILLA GÓIS, D. Comparative Analysis of the Environmental Impacts on the Dams of the Rio da Dona - Santo Antônio de Jesus-BA. In: Brazilian Symposium of Applied Physical Geography, 12, 2007. **Proceedings.** Natal, UFRN, 2007. CDROM.

NIMER, E. **Climatologia do Brasil.** 2nd edition. Rio de Janeiro. Brazilian Institute of Geography and Statistics Foundation. IBGE. 1989.

NITTA, T.; NAGAKONI, Y; SUZUKI, Y; HASEGAWA, Y; KAOKURA, A. Global analysis of the lower tropospheric disturbances in the tropies during the northern summer of IGGE year. Part I: Global features of the disturbances. **Pure Applied Geophysics,** 123, p.272-292, 1985.

NOBRE, C.A. Tropical heat sources and their associated large-scale atmospheric circulation. **Anais...** First Conference on Southern hemisphere meteorology. Preprint Volume, p. 104-107, São José dos Campos/SP, 1983.

NOBRE, P.; SHUKLA, J. Variations of sea surface temperature, wind stress and rainfall over the tropical Atlantic and South America. **Journal of Climate,** V. 9, p. 2464-2479, 1996.

OLIVEIRA, A.S. **Interactions between frontal systems in South America and convection in the Amazon.** Master's thesis. São José dos Campos, INPE. 1986.

OLIVEIRA, A. P.; FITZJARRALD, D. R. The Amazon river breeze and the local boundary layer: I - Observations. **Boundary Layer Meteorology,** 63(1-2): 141 - 162,1993.

OLIVEIRA, D.F.T.; SILVA, B.L.; DUTRA, E.M.G. East waves in 2009. XVI Brazilian Congress of Meteorology 2010. **Proceedings.** 2009.

OLIVEIRA, F.P.; OYAMA, M.D. Frequency of lines of instability and convection over the northern coast of Brazil. In: International Symposium on Climatology. João Pessoa-PB, **Proceedings,** 2011.

OLIVEIRA, L.L. et al. Study of the behaviour of rainfall at different times of the day in Belém-PA. In: **I Simpósio de Recursos Hídricos da Amazônia**, Manaus-AM, Núcleo de Hidrometeorologia, SECTAM, 2003.

OLIVEIRA, J.V.; VITORINO, M.I.; ABREU SÁ, L.D. Implication of the Madden-Julian Oscillation phase on the Eastern Amazon climate. **Atmospheric Science Letters,** v. n/a, p. n/a-n/a, 2015.

OLIVEIRA, J. V. **The Madden-Julian oscillation in eastern Amazonia: Surface variables.** Master's dissertation. 2012.

OLIVEIRA, G. B; ALCANTARA, C. R, SOUZA, E. P. Wind profile characterisation ın the formation environment of Amazonian instability lines. **Ciência e Natura,** Santa Maria v.38 n.l, Jan.- Apr. p. 393 - 403. 2016 DOI: 10.5902/2179- 460X18337

ORLANSKI, I. A rational subdivision of scales for atmospheric processes. **Buli. Amer. Meteor. Soc.,** 56 15): 527-534, 1975.

PHILANDER, S.G.H. EI Nino, La Nina, and the Southern Oscillation. **Academic Press,** 293 pp, 1990.

PLANCHON, O.; DAMATO, E; DUBREUIL, V.; GOUERY, P.; A method of identifying and locating sea-breeze fronts in north-eastern Brazil by remote sensing. **Meteorol. Appl.,** 13, 225-234, 2006.

PONTES DA SILVA, B.E. **Contribution of easterly wave disturbances to rainfall in eastern Northeast Brazil: mean synoptic evolution and numerical simulations.** Master's dissertation-Institute of Astronomy, Geophysics and Atmospheric Sciences, University of São Paulo. 2011.

QUADRO, M.RL. **Study of South Atlantic Convergence Zones (SACZ) over South America.** 97p. Master's Thesis - National Institute for Space Research, São José dos Campos. 1994.

RAMOS DA SILVA, R.; Gandu, A. W; Sá L. D; Dias, M. A. S. Cloud streets and land-water interactions in the Amazon. **Biogeochemistry** 105(1-3): 201-211, doi: 10.1007/S10533-0U-9580-4. 2011.

RAO, V.B.; HADA, K. Characteristics of rainfall over Brazil: annual variations and connections with Southern oscillation. **Theoretical and Applied Climatology,** v. 42, p. 81-91, 1990.

RAO, M.V.; PALIYATH, G.; ORMROD, D.P. Ultraviolet-B-and ozone-induced biochemical changes in antioxidant enzymes of Avabi-dopsis thaliana. **Plant Physiol,** v. 110, p. 125-136,1996.

REPELLI, C.A.; SOUZA, E.B.; QUADRO, M.F.L.; ALVES, J.M.B.; SAKAMOTO, M.S.: The Intense Rainfall Episode in the Brazilian Northeast at the End of March/1997: Influence of the 30-60 Day Oscillation. **Revista Brasileira de Meteorologia,** 13, 1-18,1998.

RICHL, H.; MALKUS, J. S. On the heat balance in the equatorial trough zone. **Geophysical.** 6, 503-538, 1958.

RIEHL, H. Waves in the easterlies and the polar front in the Tropics. Chicago University, Department of Meteorology, 79p., 1945.

RIEHL, H; SIMPSON, R. H, Climate and weather in the tropics. **Academic Press:** London, UK. 1979.

ROBLEDO, F. A.; PENALBA, O. C.; BETTOLLI, M. L. Teleconnections between tropical-extratropical oceans and the daily intensity of extreme rainfall over Argentina. **International Journal of Climatology,** v. 33, p. 735-745, 2013. DOI: 10.1002/joc.3467.

ROCHA, R. P. **Numerical simulation of mesoscale systems over South America.** São Paulo. Thesis (Master's Degree in Meteorology). Institute of Astronomy, Geophysics and Atmospheric Sciences. University of São Paulo. 1992.

RODRIGUES, D. A.; CAVALCANTI, I. F. A. Simulations of the hydrological cycle over Southern South America, Using the CPTEC/COLA AGCM. **J. Hydrometeor,** V. 7, p. 916-936, 2006.

ROMESBURG, H. C. **Cluster analysis for researchers.** Belmont, CA: Lifetime Learning Publications, 1984.

ROPELEWSKI, C. E; HALPERT, M. S. Global and regional scale precipitation associated with EI Nino/Southern Oscillation, **Mon. Weather Rev.,** 115, 1606-1626, 1987.

ROPELEWSKI, C. F.; HALPERT, M. S. Precipitation patterns associated with the high index phase of the Southern Oscillation. **J. Climate,** 2, 268-284, 1989.

ROTUNNO, R., J. A; CURRY, C. W; FAIRALL1, C. A; FRIEHE, W. A; LYONS,, J. E; OVERLAND, R. A; PIELKE, D. P; ROGERS, S. A. Coastal Meteorology, A Review of the State of the Science. **National Academy Press,** 99 pp, 1992.

SALIO, P.; NICOLINI, M.; ZIPSER, E.J. Mesoscale convective systems over southeastern South America and their relationship with the South American low-level jet. **Monthly Weather Review,** v.135, n.4, p.1290-1309, 2007.

SANTOS, O.C.O. **Analysis of land use and water resources in the Igarapé Apeú watershed, north-eastern Pará State.** Doctoral thesis. Postgraduate Programme in

Geography. Federal University of Rio de Janeiro. Rio de Janeiro, 269p., 2006.

SANTOS, A. M.; GALVÍNCIO, J. D.; MOURA, M. S. B. Homogenisation of rainfall in the Goiana River basin - PE, using the cluster analysis method. **Revista Brasileira de Geografia Física,** Recife, v. 1, n. 1, p. 14-27.2008.

SANTOS, S. R. Q.; BRAGA, C. C.; VENTURA, E. D.; CAMPOS, T. B.; SANTOS, A. P. P. Application of the Standardised Precipitation Index (SPI) to the city of Belém. In: XVII Brazilian Congress of Meteorology, Gramado - RS. In: XVII Brazilian Congress of Meteorology, Gramado-RS. **Proceedings...**Gramado:SBMet 2012, 1 CD- ROM, 2012.

SATYAMURTY, P.; COSTA, C. P. d.; MANZI, A. Moisture source for the Amazon Basin: a study of contrasting years. **Theoretical Applied Climatology,** 2012.

SETH, A.; FERNANDES, K.; CARMAGO, S.J. Two summers of São Paulo drought: Origins in the Western Tropical Pacific. **Geophysical Research Letters,** Wiley Online Library, 2015.

SCOLAR, J.; FIGUEIREDO, J.C. Analysis of the synoptic conditions associated with the formation of mesoscale convective complexes. In: Brazilian Congress of Meteorology. **Anais,** 7, Salvador: SBMET, v.2, p.455-61,1990.

SHIMIZU, M. H.; AMBRIZZI, T. MJO influence on ENSO effects in precipitation and temperature over South America. , Volume 124, , pp 291-301, 2016. DOI: 10.1007/s00704-015-1421-2.

SILVA DIAS, M. A. F.; FERREIRA, R. N. Application of a linear spectral model to study of Amazonian squall lines during GTE/ABLE 2B. **J. Geophys. Res; V.** 97, p. 20405-20419, 1992.

SILVA DIAS, M. A. F.; SILVA DIAS, P. L.; LONGO, M.; FITZJARRALD, D. R.; DENNING, A. S. River breeze circulation in eastem Amazonia: Observations and modelling results, **Theor. Appl. Climatol.,** 78,111-121, 2004.

SODRÉ, G. R. **Study of Mesoscale Convection on Different Surfaces in Eastern Amazonia.** Master's thesis - Federal University of Pará, Belém, 2013.

SODRE, G.R.C.; VITORINO, M.I. Observational analysis of mesoscale convective systems associated with climatic extremes in the eastern Amazon. **Anais.** XVI Simposio Brasileiro de Sensoriamento Remoto-SBRS, Foz do Iguaçu, PR, Brazil, 13 April 2013, INPE.

SODRE, G. R; VITORINO, M. I; COHEN, J. C. P; MORAES, B. C. Observational study of mesoscale convection on different surfaces in the state of Pará. **Revista Brasileira de Geografia Física,** v. 08 n. 05. p. 1281-1293. 2015.

SOKAL, R., & MICHENER, C. D. A statistical method for evaluating systematic relationships. **University of Kansas Science Bulletin,** 38, 1409-1438. 1958.

SOUZA, E. B.; NOBRE, P. A review of the dipole pattern in the tropical Atlantic. **Revista Brasileira de Meteorologia,** V. 13, n. 1, p. 31-44,1998.

SOUZA, E B; KAYANO, M T; TOTA, J; PEZZI, L; FISCH, G; NOBRE, C: One the influence of El Nino, La Nina and Atlantic Dipole pattern the Amazonian rainfall during 1960

- 1998, **Acta Amazonica** , v. 30, p. 305 - 318, 2000.

SOUZA, E.B.; KAYANO, M.T.; AMBRIZZI, T. Intraseasonal and submonthly variability over the Eastern Amazon and Northeast Brazil during the autumn rainy season. **Theoretical and Applied Climatology,** 81, p.177-191, 2005.

SOUSA, J. R. A.; ALMEIDA, R. M. B.; ROLIM, P. A. M. Influence of the Atlantic Dipole on precipitation in eastern Amazonia/Northern Brazil (Macapá- AP, Belém-PA and São Luis-MA). **VIII Brazilian Congress of Meteorology, Belo Horizonte/MG,** 2002.

SOUZA, E. B.; AMBRIZZI, T. Pentad precipitation climatology over Brazil and the associated atmospheric mechanisms. **Climanálise.** Available at: http://cptec.inpe.br/products/cliamanalise/artigos/artcien2.shtml/. 2003.

SOUZA, R. B.; ROBINSON, I. S. Satellite and Lagrangian observation of the Brazilian costal current. **Continental Shelf Research,** V. 24, p. 241-262, 2004.

SOUZA, A. M. L. **Observational study of low-level jets on the north and northeast coasts of Pará during the rainy and dry seasons.** 154 f. Dissertation (master's degree in meteorology) - Federal University of Pelotas, Pelotas, RS, 2005.

SOUZA, E. B.; AMBRIZZI, T.: Modulation of the intraseasonal rainfall over tropical Brazil by the Madden-Julian oscillation. **International Journal of Climatology.** DOI: 10.1002/joc. 1331, 2006.

SCOLAR, J. E FIGUEIREDO, J. C. Analysis of the synoptic conditions associated with the formation of mesoscale convective complexes. In: VI CBMet. **Proceedings,** 2: 455-461,1990.

SNEATH, P. H. A. The application of computers to taxonomy. **Journal of General Microbiology,** 17, 201-226, 1957.

STATHIS, D.; MYRONIDIS, D. Principal component analysis of precipitation in Thessaly Region (Central Greece). **Global NEST Journal,** 11(4): 467-476, 2009.

STRANG, D. M. G. **Climatological analysis of rainfall normals in northeastern Brazil.** Pap. No. IAE-M-02/72. Aerospace Technical Centre. São José dos Campos, 70 p. 1972.

SUI, C. H.; LAU, K. M.; TAKAYABU, Y. N.; SHORT, D. A. Diurnal Variations in Tropical Oceanic Cumulus Convection during TOGA COARE. **Journal of the Atmospheric Science,**
Washington, v. 54, n. 5, p. 639-655, Mar. 1997.

TEIXEIRA, R. F. B. The breeze phenomenon and its relationship with rainfall over Fortaleza-CE. **Revista Brasileira de Meteorologia,** V. 23, n. 3, 282-291, 2008.

TORRES, A. B. Rainfall distribution over the Doce River basin in Minas Gerais. In: Brazilian Symposium of Applied Physical Geography, 12. **Proceedings.** Natal, UFRN, 2007. CD-ROM, 2007.

TORRES, R. **Analysis and Modelling of East Wave Disturbances acting on the East Coast of Northeast Brazil** (INPE-15459-TDI/1421). Dissertation (Master's in Meteorology)

- National Institute for Space Research, São Jose dos Campos, 2008.

TUCCI, C. E. M. **Hidrologia Ciência e Aplicação.** Editora da UFRGS EDUSP ABRH p 952, 1993.

TUCCI, C. E.; DIAS, P.; CLARKE, R.; COLLISCHONN, W.; SAMPAIO, G. Comparison of flow forecasts by runoff models: A case study from Southern Brazil. **International Association of Hydrological Sciences,** Publication, v. 280, p. 142- 148, 2003.

TRENBERTH, K.E. The definition of EI Nino. Buli **.Amer. Met. Soc.,** 78 (12), 2771- 2777, 1997.

TRYON, R. C. **Cluster analysis.** Ann Arbor: Edwards Brothers, 1939.

UVO, C. B. **The Intertropical Convergence Zone (ITCZ) and its relationship with precipitation in the North of the Brazilian Northeast.** 1989. 99 f. dissertation (Master's in Meteorology) - INPE, São José dos Campos, 1989.

VAREJÃO-SILVA, M. A. **Meteorology and Climatology.** Digital Version 1. In: Atmospheric Disturbances, p. 444-446. Recife, 2005.

VELASCO, L; FRITSCH, J. M. Mesoscale Convective Complexes in the Americas. **J. Geophys. Res.** 92, p. 9591-9613. 1987.

VIANELLO, R.L.; ALVES, A. R. **Meteorologia e aplicações.** Viçosa: UFV, 449p, 1991.

VITORINO M. I; SILVA M. E. S; ALVES J. M. B. Classification of mesoscale convective systems in the northern sector of the Brazilian Northeast. **Revista Brasileira de Meteorologia.** 12(l):p. 21-32. 1997.

VITORINO, M. I. **Analysis of intraseasonal oscillations over South America and adjacent oceans using wavelet analysis.** 2003. 344 f. Thesis (doctorate in meteorology) - INPE. São Jose dos Campos, 2003.

VITORINO, M. L, SILVA, P. L. D.; FERREIRA, N. J. Observational study of the seasonality of the sub monthly and intraseasonal signal over the tropics. **Meteorology and Atmospheric Physics,** n. 93, p. 17-35. 2006.

WALISER, E. W. C.; GAUTIER, C. A Satellite-derived Climatology of the ITCZ. **Journal of Climate,** 6. p.2162-2174. 1993.

WARD, J. H. Hierarchical Grouping to Optimise an Objective Function. **Journal of the American Statistical Association.** n. 58, p. 236 -244, 1963.

WILKS, D.S. Statistical methods in the atmospheric Sciences, Second Edition. **International Geophysical Series.** Department of Earth and Atmospheric Sciences, Cornell University. Elsevier. v.ll, 463 p. 2006.

WISHART, D. (1969). An algorithm for hierarchical classifications. **Biometrics,** 25, 165-170.

YAMAZAKI, Y. **Theoretical and synoptic studies of tropical disturbances.** Master's dissertation. São José dos Campos, SP, INPE, 46p., 624-LAFE, 1975.

YANG, G., J. SLINGO. The diurnal cycle in the tropics. **Monthly Weather Review,** v. 129, p.784-801, 2001.

ZHANG, C. Madden-Jullian Oscillation. **Reviews of Geoohysics.** v.43, Issue 2, 1 of 36, RG 2003/2005. DOI: 10.1029/2004RG000158.

ZIPSER, E. J. The role of organised unsaturated convective downdraft in the structure and rapid decay of an equatorial disturbance. **J. Appl. Meteor,** 8, 799-814, 1969.

CHAPTER 7

ANNEX

TIME	JANUARY	FEBRUARY	MARCH	APRIL	MAY	JUNE
1	34	47	45	29	19	12
2	38	36	44	39	19	7
3	32	49	36	36	23	12
4	30	47	35	31	13	10
5	32	43	30	33	14	5
6	27	53	34	27	20	8
7	31	47	34	25	20	11
8	28	37	28	26	14	6
9	30	22	26	26	11	9
10	25	21	23	19	16	9
11	24	25	24	22	14	9
12	46	56	38	24	11	4
13	66	78	100	69	41	17
14	112	129	166	133	61	36
15	151	157	182	143	91	44
16	161	170	171	148	109	60
17	137	153	162	123	119	76
18	127	117	123	102	113	60
19	108	109	114	106	101	63
20	68	92	105	76	93	59
21	50	66	93	90	82	52
22	40	56	74	66	55	32
23	32	47	62	60	43	22
24	29	49	52	34	25	17

TIME	JULY	AUGUST	SEPTEMBER	OCTOBER	NOVEMBER	DECEMBRC
1	14	8	5	0	3	14
2	7	5	3	1	3	9
3	9	3	4	3	1	13
4	5	3	5	1	3	6
5	5	3	9	3	3	5
6	6	3	6	2	6	7
7	10	5	3	3	2	7
8	3	2	0	2	2	6
9	5	5	4	1	7	9
10	7	8	4	8	4	7
11	9	4	3	3	5	35
12	3	4	4	3	14	66
13	17	8	4	15	15	33
14	16	12	25	37	28	80
15	25	30	45	72	66	120
16	32	46	79	84	71	126
17	56	68	88	79	65	110
18	56	61	70	41	43	79
19	66	65	50	29	29	74
20	60	54	32	17	15	54
21	53	34	14	8	9	27
22	40	13	11	5	6	17
23	24	10	3	6	6	16
24	14	7	3	2	4	18

MIX
Papier aus verantwortungsvollen Quellen
Paper from responsible sources
FSC® C105338

Printed by Books on Demand GmbH, Norderstedt / Germany